RAND ARROYO CENTER

T0122917

Enhanced Army Airborne Forces

A New Joint Operational Capability

John Gordon IV, Agnes Gereben Schaefer, David A. Shlapak, Caroline Baxter, Scott Boston, Michael McGee, Todd Nichols, Elizabeth Tencza

Prepared for the United States Army

For more information on this publication, visit www.rand.org/t/rr309

Library of Congress Cataloging-in-Publication Data

Gordon, John, 1956-
  Enhanced Army Airborne Forces : a new joint operational capability / John Gordon
IV, Agnes Gereben Schaefer, David A. Shlapak, Caroline Baxter, Scott Boston, Michael
McGee, Todd Nichols, Elizabeth Tencza.
    pages cm
  Includes bibliographical references.
  ISBN 978-0-8330-8216-9 (pbk. : alk. paper)
  1. United States. Army—Airborne troops—Reorganization. 2. United States.
Army—Airborne troops—Equipment. 3. Armored vehicles, Military—United
States. 4. Airborne operations (Military science) I. Rand Corporation. II. Title.
  UD483.G67 2014
  356'.1664—dc23
                                                                      2014038619

Published by the RAND Corporation, Santa Monica, Calif.

© Copyright 2014 RAND Corporation

RAND® is a registered trademark.

Cover Image by Capt. Tom Cieslak

Preface

This report was written as part of a project entitled "Joint Operational Concepts for an Uncertain Future." The focus of the research became an effort to assess the challenges that U.S. Army airborne forces may face in the future and identify capabilities airborne forces will need to effectively address those challenges. This report summarizes threats to the current U.S. airborne force and explores the concept of an airborne light armored infantry force as a possible means to mitigate those threats. Additionally, the report examines possible vehicle options for such a concept, as well as joint requirements that the concept might generate, particularly in terms of the amount of airlift that an Army airborne unit with an increased number of vehicles could require. The research also examines potential uses for such an airborne light armored infantry force, advantages and disadvantages of the new concept, and issues related to implementation. Importantly, the research focused on near-term options (the next three to five years) to improve the capabilities of today's airborne forces.

This research was sponsored by the U.S. Army's Director of Concepts and Learning, Army Capabilities Integration Center, Training and Doctrine Command Headquarters, and was conducted in the RAND Arroyo Center's Strategy and Resources Program. RAND Arroyo Center, part of the RAND Corporation, is a federally funded research and development center sponsored by the United States Army.

The Project Unique Identification Code (PUIC) for the project that produced this document is RAN126151.

Contents

Preface iii

Figures ix

Tables xi

Summary xiii

Acknowledgments xxiii

Abbreviations xxv

CHAPTER ONE
Introduction 1

CHAPTER TWO
Overview of the Current Airborne Force 7
Historical Utilization of the Airborne Force 7
    Operation Just Cause 8
    Operation Desert Shield 9
    Operation Uphold Democracy 9
    Operation Iraqi Freedom 9
    Recent Operations in Iraq and Afghanistan 10
Review of Current Airborne Force Structure 11
Strengths and Weaknesses of Today's Airborne Force 12

CHAPTER THREE
Threats to Today's Airborne Forces 15
Improved Air Defenses 16
    Medium-/High-Altitude Defenses 16
    Low-Altitude Defenses 20

Implications of These Threats 24
Long-Range Fires Directed Against Drop Zones and Airheads 24
The Ground Threat to Airborne Operations 28

CHAPTER FOUR
A Proposed Airborne Light Armored Infantry Force 31
Overview of the Concept 31
Light Armored Vehicle Options 34
    LAV-II 34
    Stryker 35
    HMMWV 37
Airlift Requirements 38
    Stryker/LAV Brigade 39
    Stryker/LAV Battalion Task Force 39
    Drop Zone Requirements and Aircraft Operations Profiles 42

CHAPTER FIVE
Airlift Issues and Requirements 45
    Delivery Aircraft 45
    Limitations and Considerations 46
    Analysis: Stryker/LAV-Based Airborne Light Armored Infantry
        Brigade 49
    Analysis: Stryker/LAV-Based Airborne Light Armored Infantry
        Battalion Task Force 50
Summary 54

CHAPTER SIX
Potential Uses for Airborne Light Armored Infantry Forces 57
Vignette 1: Counter Genocide 58
Vignette 2: Establish a Deterrent Presence 60
Vignette 3: Protect an Enclave 61
Vignette 4: Seize and Secure a WMD Site 63
Vignette 5: Conduct a Noncombatant Evacuation Operation 65
Vignette 6: Conduct a Humanitarian Assistance and Disaster Relief
    Operation 67

Vignette 7: Airborne Light Armored Infantry Forces in
    State-to-State Conflict 68
New Role: Airborne Cavalry... 69
Concluding Thoughts 70

CHAPTER SEVEN
Issues Related to the Implementation of the New Concept.............. 73
Issues the Army Would Need to Resolve..................................... 73
Implications of This Concept for the Joint Force 74
Organizational Options for Implementation 75

CHAPTER EIGHT
Conclusions and Recommendations 79
Recommendations ,, 79

APPENDIXES
A. LAV-II Family of Vehicles 83
B. Stryker- and LAV-Based Airborne Light Armored Infantry
    Brigade TOEs... 89
C. C-5, C-17, and C-130 Capabilities..................................... 97
D. Dimensions, Weight, Number of Vehicles for C-17 Airdrop 99

Bibliography... 103

Figures

3.1.	SA-15	17
3.2.	S-60 57-mm Anti-Aircraft Gun	20
3.3.	SA-18	22
3.4.	Fajr-5	27
5.1.	C-17 Range Versus Load	48
6.1.	Yongbyon Concept	64
6.2.	Nigeria	66
A.1.	LAV-25	85
A.2.	LAV-IIH	85
A.3.	LAV-AG	87
A.4.	LAV-AT	87
A.5.	LAV-M	88

Tables

S.1. C-17s Required for Stryker- and LAV-Based Airborne
Light Armored Infantry Battalion Task Forces xix
S.2. Potential Advantages and Disadvantages of New
Airborne Concepts... xvii
4.1. Airborne Light Armored Infantry Brigade Airdrop
Echelons... 40
4.2. Airborne Light Armored Infantry Battalion Task Force........ 41
4.3. Vehicular Weapons of a LAV-II Force Compared with a
Stryker Force 42
5.1. Stryker-Based Airborne Light Armored Infantry Brigade
Airdrop Echelons 51
5.2. LAV-Based Airborne Light Armored Infantry Brigade
Airdrop Echelons 52
5.3. C-17 Sorties Required for Stryker and LAV-Based
Airborne Light Armored Infantry Battalion Task Forces........ 53
5.4. C-17 Sortie Requirement for Stryker and LAV Brigades........ 54
8.1. Potential Advantages and Disadvantages of New
Airborne Concepts.. 82
B.1. First- and Second-Echelon TOEs............................. 90
B.2. Third- and Fourth-Echelon TOEs............................. 93
C.1. C-5, C-17, and C-130 Lift Capabilities 98
C.2. C-5 Airlift Options.. 98
D.1. C-17 Airdrop Scenario Characteristics...................... 100

Summary

This project focuses on the pivotal role that airborne forces could play in key missions in the future—particularly against hybrid threats and in anti-access environments. Army airborne forces are unique in their ability to quickly deploy worldwide from the continental United States via transport aircraft, including to objectives that may be deep inland and generally beyond the reach of maritime forces operating in littoral regions. However, the threats facing Army airborne forces today (and the Air Force transport planes that deploy and sustain them) are serious and could become more severe in the future, depending on the opponent. For instance, in the future, airborne forces will likely be confronted with increasingly sophisticated anti-access threats, including evolving low-, medium-, and high-altitude air defense systems; longer-range surface-to-surface fires threatening the lodgments that are critical to the success of airborne operations; and improved tactical combat capabilities in the hands of potential opponents. To overcome these new threats, the airborne force will need new capabilities.

A very important consideration in this research was the need to identify potential enhancements to today's airborne forces that could be made in the next three to five years. That consideration had the effect of limiting material solutions to essentially extant equipment. Additionally, the near-term focus also meant that there would be no significant increases to the Air Force's airlift fleet during this time frame.

The research team identified the concept of enhancing today's airborne forces by adding a light armored infantry capability as one potential response to these threats. Such a capability that could be air-dropped by parachute or air-landed at an airfield could provide airborne forces with increased speed and mobility once it arrived in the objective area, as well as greater survivability and firepower. This greater level of mobility would have both tactical and operational implications, giving the more heavily armed and better-protected airborne force the ability to maneuver over larger areas.

We conducted extensive analyses of vehicle options for this concept, as well as the airlift requirements that such a concept would generate. In addition, we conducted a tabletop exercise with subject-matter experts to explore how such a concept could be employed in a variety of scenarios.

Today's Airborne Force

As of late 2012, the Army has six infantry brigade combat teams (BCTs) organized and equipped for this mission: the four BCTs of the 82nd Airborne Division at Fort Bragg, the 173rd Airborne Brigade Combat Team in Italy, and the 4th Infantry Brigade Combat Team (Airborne), 25th Infantry Division in Alaska. In addition, the XVIII Airborne Corps has a number of combat support elements also capable of carrying out airborne operations, including engineer and sustainment units. The units mentioned here do not include special operations forces that are trained to carry out airborne operations, including the 75th Ranger Regiment and Army special forces.

Today's airborne forces lack protected armored mobility and the ability to conduct mounted combat, and they have limited tactical mobility. From the 1960s to the early 1990s, the 82nd Airborne Division included a battalion of M551 Sheridan light tanks that could be airdropped. When that battalion was disbanded, there was no replacement for that capability.

Threats to Today's Airborne Forces

There are two general classes of air defenses that influence airborne operations: medium-/high-altitude defenses, and low-altitude defenses. Medium-/high-altitude defenses generally apply to aircraft flying above 15,000 feet. The very high-quality Russian-built 9K330 Tor system (North Atlantic Treaty Organization [NATO] designation SA-15) is an example of this class of radar-guided surface-to-air missile (SAM) capability. With a maximum altitude of more than 20,000 feet and a range of roughly 12 km from launcher to target, the SA-15 is highly resistant to electronic countermeasures, can track multiple targets simultaneously, and is able to fire while on the move; individual Tor launcher vehicles can operate independently if necessary. Other, larger, radar-guided SAMs have much longer ranges. The Russian-produced S-300/400 series SAMs (NATO designations SA-10/20/21) are much larger than the SA-15 and have ranges of up to 400 km, depending on the specific model of missile.

The other class of air defenses that can affect airborne operations is the low-altitude threat. The nature of this challenge is very different from the medium/high-altitude threat. Whereas weapons such as SA-15 or SA-10/20 must turn on their radars to acquire, track, and engage high-flying aircraft, that is not the norm for low-altitude defenses. Anti-aircraft artillery (AAA) and man-portable air defense systems (MANPADS, shoulder-fired missiles) are the most important low-altitude threats.

The proliferation of long-range, surface-to-surface indirect fires systems also poses a growing threat to airborne operations. For instance, many airborne operations are initiated with a parachute operation to seize a lodgment for follow-on forces to arrive. During these parachute operations, personnel and equipment may be vulnerable to long-range surface-to-surface systems. Airborne operations often include seizing an airfield for follow-on forces' arrival. Personnel and equipment may also be vulnerable to surface-to-surface systems during this type of operation.

The Russian BM-30 Smerch is an example of the type of long-range surface-to-surface system available today. Today the 12-barrel

Smerch has a maximum range of roughly 90 km—and that is being increased to roughly 120 km. Its 300-mm rockets can carry a mix of warheads, including submunitions. Submunitions delivered from long range are a particularly serious threat to an airfield, since a single battery of BM-30 launchers could saturate a runway and aircraft parking area in a single salvo. Transport planes on the ground are highly vulnerable to explosions and fragmentation. Even if no aircraft were hit, if the runway were littered with small fragments from exploded submunitions, air operations would be shut down until the runways were cleared. Even with a dedicated runway repair team, this could take several hours. Importantly, the Air Force does not have an airdroppable engineering capability for runway clearing and repair. Army airborne units would have to provide that capability until Air Force runway repair units arrived.

A New Operational Concept: Light Armor for Airborne Infantry Forces

The concept developed by the RAND research team focuses on introducing light armored vehicles into today's airborne force to mitigate many of the threats described here. A very important consideration when developing options for Army airborne forces is the size and composition of the Air Force's transport aircraft fleet. The C-5 (including the C-5M, which recently began to enter service) has a very large payload capacity (over 100 tons) but requires a long runway. The C-17 Globemaster has a maximum cargo capacity of roughly 80 tons (the planning factor for the C-17 is 45 tons), but as of late 2012, not all Air Force C-17 crews were qualified for parachute operations. The C-130 is the most numerous but smallest Air Force transport plane, which means that there are important limits on the payload of the aircraft for air-landing or airdrop. However, all C-130 crews (whether active or reserve) are airdrop-qualified.

Key elements of the concept include the following:

- A suite of light armored vehicles for a variety of roles, such as assault guns (light tanks), armored personnel carriers, mortar carriers, command-and-control vehicles, ambulances, reconnaissance vehicles, and anti-armor vehicles. The vehicles would be able to fight on the move and provide excellent tactical mobility and limited armor protection for crews and passengers.
- All or most of these vehicles would be capable of parachute drop from existing Air Force transport planes. Ideally, as many of the vehicles as possible would be capable of deploying by C-130 (including by airdrop) taking advantage of the large number of C-130s that are qualified for parachute operations.
- Airborne forces equipped with these vehicles would be able to select drop zones outside the worst of the enemy's air defenses and surface-to-surface fires, then use the mobility, firepower, and protection of the vehicles to maneuver toward their objectives. The enhanced airborne force we envision would have improved tactical and operational maneuver potential.
- The maneuvering airborne units equipped in this manner would rely heavily on joint fires (e.g., close air support) and intelligence, surveillance, and reconnaissance systems, such as unmanned aerial systems for situational awareness.
- As the airborne light armored infantry force maneuvers toward its objective from drop zones that would be farther from the objective than is the case today, it would rely on aerial resupply from Air Force transport planes, delivered accurately from medium altitude via the Joint Precision Airdrop System (JPADS). Today, JPADS can deliver up to 10,000 lbs of cargo on a single pallet. (A C-17 can carry up to 11 JPADS pallets.) Accuracy is measured in tens of meters from the intended delivery point. This technique would allow Air Force transport planes to remain well above AAA and MANPADS range.
- An airhead would almost certainly still be required for the arrival of air-landing follow-on forces and portions of the required resupply, but due to the protected mobility of the airborne units, the

airfield(s) selected for seizure could be farther from the objective than is the norm in today's airborne operations.

Light Armored Vehicle Options

Given the near-term focus of this project and, therefore, the need to consider extant equipment, the Light Armored Vehicle, second generation (LAV-II), family of vehicles appeared to be the most promising candidate platform for enabling an airborne force to conduct mounted maneuver. The LAV-II series offers advantages in two broad categories. First, it is suitable as an airborne combat vehicle: It can be air-dropped from the C-17 Globemaster III, and some of the variants can be airdropped from the C-130. For a vehicle in its weight class, it has good all-around protection (14.5 mm on the frontal arc, 7.62 mm all around), firepower, and mobility. Second, the LAV-II is also available from a U.S.-aligned vendor; it is currently fielded in the U.S. Marine Corps (as the LAV-25A2 and variants) and would be available for use in short order, rather than requiring a full, new procurement effort.

The research team also considered the Stryker-series vehicles currently operated by the U.S. Army. If airdrop is an important consideration for the light armor of airborne forces, the Stryker (a derivative of the LAV-III) has significant limitations due to its weight and size. It does have the advantage of already being in the U.S. Army inventory in large numbers, though our research noted that if the Army wanted to use the Stryker in the role envisioned in this concept, some additional variants, such as vehicles armed with 25-mm guns, would probably have to be procured.

Additionally, we noted the possibility of increasing the mobility of most airborne infantry units via light, unarmored vehicles, such as the High-Mobility Multipurpose Wheeled Vehicle. In that case, an increased percentage of the infantry in airborne units would be motorized with unarmored vehicles and would operate in conjunction with new airdroppable light armor of the type described in this report.

Table S.1 shows the key systems in a Stryker or LAV-II–based battalion-sized task force, as well as the number of C-17s required to deploy the unit in a parachute configuration. In Table S.1, the columns titled "Excursion Case" show the number of aircraft required if unit personnel can ride on the same aircraft as the vehicles.

Table S.1
C-17s Required for Stryker- and LAV-Based Airborne Light Armored Infantry Battalion Task Forces

Items	Number	Stryker-Based Battalion Task Force			LAV-Based Battalion Task Force		
		Base Case	Excursion Case	Base Compared with Excursion	Base Case	Excursion Case	Base Compared with Excursion
Crew	302	3	0	−3	3	0	−3
Dismounts	384	4	0	−4	4	0	−4
Command vehicles	3	2	2		1	1	
Personnel carriers	37	18	18		12	12	
Ambulances	8	4	4		3	3	
Reconnaissance/ 25-mm vehicles	25	12	12		8	8	
Mortars	12	6	6		4	4	
Fire support vehicles	4	2	2		1	1	
Assault guns	9	5	5		3	3	
Engineer vehicles	4	2	2		2	2	
Antitank vehicles	3	2	2		1	1	
Total C-17s required		60	53	−7	42	35	−7

Recommendations

There is a need for the Army to first determine whether the light armored infantry concept is right for the airborne force. If this direction appears to be an appropriate one, the next steps should include the following:

- Refine the operational concepts associated with a new airborne light armored infantry capability. This would include a detailed examination of how such a new capability would be employed and what key joint enablers would be necessary for this mode of operations.
- Establish an experimental program. This could include obtaining LAV-II–class vehicles from the Marine Corps and elsewhere to examine their suitability for airdrop and air-landing operations. Additionally, the Stryker should undergo a detailed assessment to compare it to the LAV-II series and determine its applicability for the airborne mission.
- Examine other vehicle options. This could include a determination of whether there are any other readily available U.S. or foreign vehicles that might be appropriate.
- Determine the Air Force's main constraints to operationalize some portion of this concept. For example, the Army might require that more C-17 aircrews be qualified for airdrop operations than is the case today. The Air Force could not make such a change on short notice. Time and resources would be involved, and the Army and Air Force would need to discuss what was possible based on the Army's desired timelines and the amount of the airborne force that it would want to convert to this configuration.
- Identify additional rigging and other administrative requirements from the Army's perspective. For example, while there is currently significant rigging capability at Fort Bragg–Pope Air Force Base in North Carolina, a new requirement to rig several dozen light armored vehicles for rapid deployment could impose a burden beyond the capacity of the current rigging system. For Italy- and Alaska-based airborne units, new rigging capacity or other infra-

structure might be needed to accommodate light armored vehicles of the LAV-II or Stryker class.

- Establish the costs for the various vehicle options and the associated new units (i.e., more rigging capability and maintenance for light armor in airborne units).
- Decide on an initial organizational construct. For instance, does the Army want to convert all airborne brigades, one brigade, or just a single battalion to this configuration?

There are a number of advantages and disadvantages associated with this concept. Table S.2 summarizes key aspects of the concept for future Army airborne forces and highlights selected advantages and disadvantages.

An important aspect of what might come next is the potential usefulness of this enhanced airborne capability to combatant commanders. At present, when theater commanders and their staffs consider incorporating Army airborne forces into their contingency plans, they are basing their decisions on today's airborne forces. Some version of the concept presented in this document would be a new capability. Therefore, the Army should solicit the input of the joint headquarters that would be the ultimate customers and users of this new kind of Army capability.

Table S.2
Potential Advantages and Disadvantages of New Airborne Concepts

Key Aspect of the Enhanced Airborne Concept	Advantages	Disadvantages
Enhanced mobility, protection, and firepower of airborne units	Increases strategic and operational options for airborne forces Tactical flexibility improved	More airlift required to deploy airborne units with larger numbers of vehicles Cost of procuring light armored vehicles for the airborne force
Battalion-sized airborne light armored infantry units	Battalions easier to deploy via existing Air Force airlift assets Fewer vehicles need to be purchased compared to brigade-sized units	Battalion-sized units would probably be able to maintain only company-sized elements on high readiness Limited overall combat power
Brigade-sized airborne light armored infantry units	More combat power than battalions Able to maintain a full battalion on high level of readiness	Would require considerable airlift to deploy Higher cost due to larger number of light armored vehicles that would have to be procured
LAV-II family of vehicles	Well suited to airdrop and transport due to weight and size (low-velocity aerial delivery [LVAD]–compliant), including C-130 Family of vehicles already exists, including in U.S. military use Some compatibility with Stryker (same manufacturer)	Not currently an Army system Still-to-be-determined number of vehicles would have to be procured
Stryker family of vehicles	Currently in U.S. Army use Family of vehicles already exists	Currently Stryker is beyond weight limit of the LVAD system Difficult to transport in C-130 and cannot be dropped from it Additional vehicle types (e.g., 25 mm) would have to be procured
Use of the current Air Force airlift fleet	No new aircraft purchases needed Air Force familiar with current Army airborne concepts	Additional C-17 aircrew may need to be qualified for armored vehicle airdrop Other elements of the joint force also require airlift, especially from the C-17

Acknowledgments

The authors wish to extend thanks to the sponsors at U.S. Army Training and Doctrine Command (TRADOC) Headquarters for their support for the research, especially MG Donahue and MG Hix. In addition, LTC Mike Whetstone and Elrin Hundley at the Army Capabilities Integration Center were very helpful in providing TRADOC oversight of the effort. Lt Col James R. Twiford, AF/A5RM, Maj Scott McKeever, SAF/IARA, and Maj Joseph J. O'Rourke, AF/A8XS, offered helpful input and participated in a war game on new airborne concepts held in RAND's Arlington, Va. office on July 24, 2012. Richard Benney, the division leader of the Aerial Delivery Equipment and Systems Division of the Warfighter Protection and Aerial Delivery Directorate at the Natick Soldier Research, Development, and Engineering Center provided considerable assistance with information on current and future Army airdrop systems. Matthew Koneda, the chief engineer at the Program Manager's Office for Light Armored Vehicles helped the RAND research team assemble information on the LAV vehicle series. From the Army's XVIII Airborne Corps at Fort Bragg, North Carolina, Jon Chase and Lane Toomey provided useful insights that contributed to the research. In addition, Adam Mount conducted key analyses for the project during his tenure as a RAND summer associate. Finally, RAND Army research fellows LTC Michael Franco and LTC Chris Springer both contributed to the work, providing expertise from their recent experience in Army airborne units. MG (ret.) Paul Eaton and RAND researchers Alan Vick, Brian Nichiporuk, and Anthony Rosello provided helpful reviews and insight.

Abbreviations

AAA	anti-aircraft artillery
APC	armored personnel carrier
BCT	brigade combat team
DRB	division ready brigade
DVH	dual-v hull
FY	fiscal year
GRF	global reaction force
HADR	humanitarian assistance and disaster relief
HEMTT	Heavy Expanded Mobility Tactical Truck
HMMWV	High-Mobility Multipurpose Wheeled Vehicle
ICV	infantry carrier vehicle
IED	improvised explosive device
ISR	intelligence, surveillance, and reconnaissance
JPADS	Joint Precision Airdrop System
LAV	Light Armored Vehicle
LVAD	low-velocity aerial delivery

MANPADS	man-portable air defense system
MGS	mobile gun system
MRL	multiple-rocket launcher
MTV	medium tactical vehicle
NATO	North Atlantic Treaty Organization
NEO	noncombatant evacuation operation
NM	nautical miles
RPG	rocket propelled grenade
SAM	surface-to-air missile
SBCT	Stryker brigade combat team
TOE	table of organization and equipment
TTPs	tactics, techniques, and procedures
VRS	Vojska Republike Srpske [Army of Republike Srpska]
WMD	weapons of mass destruction
WWII	World War II

CHAPTER ONE

Introduction

As the conflicts in Iraq and Afghanistan come to an end, a new strategic vision for the Army of the future has begun to take shape. This view was made explicit in February 2011 by then–Secretary of Defense Robert M. Gates in a speech at the United States Military Academy:

> Looking ahead, though, in the competition for tight defense dollars within and between the services, the Army also must confront the reality that the most plausible, high-end scenarios for the U.S. military are primarily naval and air engagements—whether in Asia, the Persian Gulf, or elsewhere. The strategic rationale for swift-moving expeditionary forces, be they Army or Marines, airborne infantry or special operations, is self-evident given the likelihood of counterterrorism, rapid reaction, disaster response, or stability or security force assistance missions.[1]

President Obama further solidified this view in January 2012 during a speech in which he outlined the latest comprehensive defense review, *Sustaining U.S. Global Leadership: Priorities for the 21st Century*:

> As we look beyond the wars in Iraq and Afghanistan—and the end of long-term nation-building with large military footprints—we'll be able to ensure our security with smaller conventional ground forces. We'll continue to get rid of outdated Cold War–era systems so that we can invest in the capabilities that we need

---

[1] Secretary of Defense Robert M. Gates, speech delivered at the United States Military Academy, West Point, New York, February 25, 2011.

for the future, including intelligence, surveillance and reconnaissance, counterterrorism, countering weapons of mass destruction and the ability to operate in environments where adversaries try to deny us access.[2]

In response to this new strategic vision, this study's original objective was to identify (1) categories of operational challenges that pose especially serious security risks for the United States and its allies, (2) Army and joint operational concepts to address those challenges, and (3) the ground power contributions to those concepts. A few months into our analysis, the study's sponsor—U.S. Army Training and Doctrine Command (TRADOC)—asked the research team to narrow the focus of its analysis to identify (1) the role that the airborne force will likely play in the future, (2) challenges that the airborne force would likely face in the future, and (3) the capabilities the airborne force will need to effectively address those challenges.

This change in project focus reflects the pivotal role that airborne forces could play in key missions in the future—particularly against hybrid threats and in anti-access environments. Army airborne forces are unique in their ability to quickly deploy worldwide from the continental United States via transport aircraft, including to objectives that may be deep inland and generally beyond the reach of maritime forces operating in the littoral regions. However, the threats facing U.S. Army airborne forces today—and the U.S. Air Force transport planes that deploy and sustain airborne units—are serious and could become more severe in the future. Airborne forces will likely be confronted with increasingly sophisticated anti-access threats, including evolving low-, medium-, and high-altitude air defense systems; longer-range surface-to-surface fires threatening the lodgments that are critical to the success of airborne operations; and improved tactical combat capabilities in the hands of potential opponents. To overcome these new threats, the airborne force will need new capabilities.

---

[2]   President Barack Obama, "Remarks by the President on the Defense Strategic Review," transcript, January 5, 2012.

A very important consideration in this research was the need to identify potential enhancements to today's airborne forces that could be made in the next three to five years. That consideration had the effect of limiting materiel solutions to essentially extant equipment. Additionally, the near-term focus also meant that the research team would assume there would be no significant increases to the Air Force airlift fleet.

The research team identified a concept to enhance today's airborne forces by adding a light armored infantry capability as one potential response to these threats. A light armored infantry capability that could be airdropped by parachute or air-landed at an airfield could provide airborne forces with increased speed and mobility, as well as greater survivability and firepower. We conducted an extensive analysis of vehicle options for this concept, as well as the airlift requirements that such a concept would generate. In addition, we conducted a table-top exercise with subject-matter experts to explore how such a concept could be employed in a variety of scenarios.

This new concept could provide the National Command Authority with new strategic options to address potential future conflicts. It could also allow for the rapid deployment of airborne forces to address a broader set of priority missions, including many identified in the 2010 National Security Strategy, the 2010 Quadrennial Defense Review, and the 2012 Defense Strategic Guidance, such as

- conducting forcible-entry operations
- countering anti-access/area threats denial in hybrid environments
- countering or securing weapons of mass destruction (WMD)
- countering terrorism or insurgencies
- establishing a lodgment for follow-on forces
- conducting noncombatant evacuation operations (NEOs), with support from air assets
- conducting complex humanitarian assistance and disaster relief (HADR) operations
- rapidly establishing an enclave (e.g., to prevent genocide)
- rapidly interposing a peacekeeping force in a time-sensitive situation.

Enhanced airborne capabilities could also offer decisionmakers better options to stabilize potential conflicts more quickly and prevent them from escalating. Today's airborne forces can conduct many or most of the missions listed above. The types of enhancements to Army airborne capabilities described in this report could significantly improve the Army's ability to perform these missions alongside other elements of the joint force.

The near-term focus of the research meant that any proposed enhancement to Army airborne forces had to be relatively inexpensive. For example, the size and composition of the airlift fleet will not change significantly the next three to five years. Therefore, we did not assume that significant additional spending would be required by the Air Force—the concept assumed the same numbers of transport planes as exist today. Similarly, we did not envision any significant research and development plan for new armored fighting vehicles for airborne units. The near-term focus meant that existing vehicles (perhaps with minor modifications) would have to be used, as opposed to a new, expensive specialized vehicle program. As we explain in this report, enhanced airborne capabilities come with costs and trade-offs.

This report is structured as follows:

- Chapter Two provides an overview of the current Army airborne force. It examines historical cases in which the airborne force has been utilized, reviews the current structure of the airborne force, and assesses the strengths and weaknesses of today's airborne force.
- Chapter Three identifies threats to today's airborne force, including improved air defenses, longer range ground threats to airheads and drop zones, proliferation of precision-guided munitions and armored vehicles, and hybrid threats.
- Chapter Four lays out a proposed concept for an airborne light armored infantry force to address the threats identified in Chapter Three. This chapter identifies the vehicle options for such an airborne light armored infantry force.
- Chapter Five assesses the airlift requirements that would be needed to support and sustain such a force.

- Chapter Six identifies possible uses for the proposed airborne light armor force, including conventional combat against state opponents, combat against hybrid opponents, and securing WMD.
- Chapter Seven identifies issues related to the implementation of the proposed concept for an airborne light armored infantry force, including issues the Army would have to overcome, implications of this concept for the joint force, and organizational options for implementing such a force.
- Chapter Eight presents our conclusions and recommendations.

The report also includes four appendixes providing supporting data and background on the vehicles and scenarios discussed here.

This analysis is intended to assist Army decisionmakers as they consider the role of the airborne force, the missions the force may be called upon to perform, and the capabilities that it will need to successfully carry out those missions. The analysis focuses on one potential means to enhance the airborne force: incorporating light armored vehicles into the force to increase speed, mobility, and survivability. We address the risks and benefits associated with such a concept, as well as options for how such a concept could be implemented incrementally over time.

CHAPTER TWO

Overview of the Current Airborne Force

Historical Utilization of the Airborne Force

Since the creation of dedicated division-sized formations for airborne assaults in World War II (WWII), the Army has maintained the capability to conduct large airborne operations. Airborne units, and the 82nd Airborne Division in particular, have been used numerous times from the closing days of the Cold War to the mid-2000s. Some of the key operations were as follows:

- *Operation Just Cause, 1989.* The invasion of Panama in December 1989 included several airborne insertions of both special and general-purpose forces. The 82nd Airborne Division's Division Ready Brigade (DRB) carried out combat drops in Panama and transitioned into offensive air assault operations from the drop zone.
- *Operation Desert Shield, 1990.* The 82nd Airborne's DRB was deployed as a tripwire force to Saudi Arabia to deter the Iraqi Army from attempting to move south and seize Saudi oilfields.
- *Operation Uphold Democracy, 1994.* Elements of the 82nd Airborne were en route to Haiti and prepared to conduct a parachute assault when the Cédras regime agreed to relinquish power and step down peacefully.
- *Operation Iraqi Freedom, 2003.* The 173rd Airborne Brigade dropped into northern Iraq in March 2003 to establish a northern front in the absence of the 4th Infantry Division, which had been denied permission to transit through Turkey. A significant portion of the 82nd Airborne Division also participated in the

invasion in southern Iraq and was prepared to conduct parachute assaults if the need and opportunity arose.

Each case demonstrated the value of quickly deployable forces to national decisionmakers, but some cases also outlined the limitations of primarily foot-mobile, light infantry–based airborne units. For instance, in the case of the 173rd Airborne in northern Iraq in 2003, there was little desire to advance southward from the lodgment airfield toward the main Iraqi forces until heavy armor had arrived from Germany. The threat posed by low-altitude air defenses in southern Iraq made senior decisionmakers unwilling to conduct either airborne or air assault operations, since the arriving infantry would have to land close to its objectives due to the lack of tactical mobility.[1] In addition, some of the capabilities developed to support airborne forces no longer exist, such as the battalion of Sheridan light tanks that was retired from the 82nd Airborne Division in the early 1990s. We examine these historical cases in turn.

Operation Just Cause

Airborne units were used extensively in Panama, and the 82nd Airborne's 1st Brigade and the 75th Ranger Regiment were both employed in airborne insertions. Parachute assaults were used to seize airheads and quickly secure over two dozen Panamanian Defense Force targets. Airborne troops carried out air assaults via helicopters from their drop zones. The 82nd Airborne's M551 Sheridan light tanks were dropped into Panama. This was the only time in history that this capability was used in combat.[2]

The case of Panama shows the agility and responsiveness of airborne forces, particularly given a high state of readiness, as was the case at the time. However, the outcome was never in doubt. Panamanian forces were hopelessly outmatched by the U.S. military, which

---

[1]   Interviews with 173rd Airborne Brigade staff, Vicenza, Italy, 2005.

[2]   MG James H. Johnson, Jr., interview with Robert K. Wright, Jr., historian, XVIII Airborne Corps, at Headquarters, 82nd Airborne Division, Fort Bragg, N.C., March 5, 1990.

was striking at targets simultaneously throughout that tiny country from a base inside Panama itself.

Operation Desert Shield

In Operation Desert Shield, the 82nd Airborne's DRB (then the 2nd Brigade) deployed to Saudi Arabia in response to the Iraqi invasion of Kuwait. Leading the rest of the division, 2nd Brigade deployed by air-landing at Saudi airfields. They served as a tripwire force, essentially a signal to Iraq of U.S. commitment, but would have been significantly outmatched in combat power had Iraqi armored forces attacked into Saudi Arabia.[3] This deployment demonstrated the risks associated with deploying a largely foot-mobile force into a region with terrain that is ideal for the rapid maneuver of armored forces. Although the Iraqis did not advance into northern Saudi Arabia, the risks of deploying a brigade of paratroops to serve as a deterrent force were clear.

During Operation Desert Storm, the 82nd participated as part of XVIII Airborne Corps' effort to secure the western flank of the heavy armored "left-hook" ground offensive against Iraqi forces. It undertook no airborne operations in that campaign.

Operation Uphold Democracy

The 82nd Airborne was employed in 1994 during attempts to negotiate the restoration of the democratically elected Aristide government in Haiti. C-130s with elements of the 82nd were en route to conduct combat drops in Haiti when the Cédras government agreed to surrender power peacefully. Confirmation that combat troops were in the air was a key factor prompting this decision.

Operation Iraqi Freedom

The invasion of Iraq in 2003 saw the use of the 82nd Airborne Division headquarters, some divisional assets, and one brigade. While plan-

---

[3] This has been discussed at some length in previous RAND work, including John Matsumura, Randall Steeb, John Gordon, Thomas J. Herbert, Russell W. Glenn, and Paul Steinberg, *Lightning Over Water: Sharpening America's Light Forces for Rapid Reaction Missions*, Santa Monica, Calif.: RAND Corporation, MR-1196-A/OSD, 2000.

ners had at one point envisioned a possible *coup de main* through an airborne assault on Baghdad International Airport, the 82nd was used primarily as light infantry, clearing bypassed enemy formations in towns and protecting 3rd Infantry Division's lines of communication from Saddam Fedayeen and other irregulars.

In Northern Iraq, the 173rd Airborne Brigade was airdropped near Bashur airfield in March 2003 to help create a northern front as part of Joint Special Operations Task Force–North. Special forces and Kurdish *peshmerga* controlled the area, but the paratroopers were still dropped rather than assault-landed to reduce the risk to aircraft and the time needed to deploy the entire force on the ground. Additionally, the airfield runway was known to be in a state of disrepair and might have started to crack under the weight of landing cargo planes. Airdropping the initial elements of the brigade "saved" several dozen landings and likely lengthened the lifespan of the runway. The airborne troops secured the airfield and awaited the arrival of a company team from the 1st Infantry Division (Mechanized), including five Abrams main battle tanks and five Bradley infantry fighting vehicles.[4] With 17 C-17s allotted for the initial drop and 12 available for follow-on transport, the 173rd was able to deploy the brigade task force of 2,200 soldiers and more than 400 vehicles, including its small armored supporting force, in 62 C-17 sorties over four days.[5]

Recent Operations in Iraq and Afghanistan

Each of the Army's airborne brigade combat teams conducted multiple combat tours during the period of U.S. military involvement in Iraq from 2003 to 2011 and in Afghanistan beginning in 2002. For a time, the 82nd's traditional mission of providing a global response capability was given to a brigade of the 101st Airborne Division, as all of the 82nd's brigade combat teams (BCTs) were engaged in deployment

---

[4]   Jamie L. Krump, "Sustaining Northern Iraq," *Army Sustainment*, November–December 2003.

[5]   Thomas W. Collins, "Parachute Assault Demonstrates Army's Strategic Responsiveness: 173rd Airborne Brigade in Iraq," *Army Magazine*, June 2003.

rotation cycles. The Global Response Force (GRF) role has been resurrected and has been integrated into the Army Force Generation cycle.

Taken as a whole, the current airborne force is in a state of reset and transition, combining recent combat experience and a new BCT-centric organization with the need to regain some of the tactical knowledge and expertise of the pre–Iraq War era. As the Army continues to restore the 82nd's ability to perform as a premier strategic response force, and for the independent airborne brigades to perform as regional first responders, it is an appropriate time to consider what changes and enhancements would be most useful in ensuring that airborne forces provide the greatest utility possible to the nation's leadership.

Review of Current Airborne Force Structure

As of late 2012, the Army has six airborne infantry BCTs organized and equipped for this mission: the four 82nd Airborne BCTs at Fort Bragg, the 173rd Airborne BCT in Italy, and the 4th Infantry BCT (Airborne), 25th Infantry Division in Alaska. In addition, XVIII Airborne Corps has a number of elements capable of carrying out airborne operations, including engineer and sustainment units. The units mentioned here do not include special operations forces that are trained to carry out airborne operations, including the 75th Ranger Regiment and Army special forces.

The changes made to transition the Army into a force comprising modular brigade combat teams affected all the Army's airborne units, particularly the 82nd. The 82nd Airborne as a division transitioned from three brigades with a total of nine parachute infantry battalions to four brigades with eight infantry battalions.[6] The 82nd Airborne Division transitioned from a division-based structure to one based around modular brigade combat teams; the 173rd Airborne Brigade transitioned into an Airborne BCT, and 4th BCT (Airborne) was cre-

---

[6] While no official announcement has been made, early indications are that the 82nd will transition back to a three-brigade structure, with nine infantry battalions, though additional details are not currently available.

ated with a cadre from an airborne battalion task force at Fort Richardson, Alaska. Each BCT, as currently organized, has a reconnaissance squadron and an artillery battalion, as well as engineers and other key enablers at the brigade level. Additionally, the 82nd Airborne Division has the 18th Fires Brigade, which can deploy 155-mm towed howitzers via airdrop and the High Mobility Artillery Rocket System by air landing.

Strengths and Weaknesses of Today's Airborne Force

The Army has fielded division-sized airborne units since August 1942, when the 82nd Infantry Division was reclassified as a parachute unit. Today the Army's airborne capability provides it with a unique ability to rapidly deploy up to brigade-sized formations from Fort Bragg, northern Italy, or Alaska for short-notice operations anywhere in the world.

An advantage of today's airborne capability is its ability to conduct forced entry operations into areas that are deep inland. Whereas maritime forces can reasonably reach a few hundred miles inland if they are located near a crisis, airborne forces can be inserted virtually anywhere, so long as sufficient transport planes and tanker support are available and if the threat can be managed to permit airborne operations. This is an important capability in an era when Army missions could include the need to rapidly secure WMD or rush to protect U.S. citizens.

While they provide a unique capability to deploy globally within a few hours or days, today's airborne forces face emerging challenges from area-denial weapons that limit their utility. The next chapter elaborates on those threats, which apply both to the Army's airborne units once on the ground in the operational area and to the Air Force transport aircraft that are used to carry the airborne force. The key trade-off that has confronted airborne forces since World War II still applies today: The more equipment added to airborne units to increase their capability when they arrive in the operational area, the greater the number of aircraft required to deploy and sustain that force. Certain types of equipment can also constrain the specific type of transport plane that can be used. This is a constant issue with the C-130 Hercules tactical airlifter, which is limited to no more than a 20-ton payload (and less if the load is to be airdropped).

Two important factors that constrain today's airborne force are: (1) the rather limited tactical mobility of airborne units on the ground, thus requiring the seizure of airheads relatively close to the objective area (areas that the opponent will probably defend), and (2) the lack of organic armored support. When the M551 Sheridan was retired in the early 1990s, the M8 Armored Gun System was envisioned as its replacement. However, the M8 was canceled in 1996, and no follow-on vehicle was developed. Although the 105-mm Mobile Gun System variant of the Stryker can be air-landed by C-17 or C-5 aircraft, it cannot be carried by C-130 and is too heavy for the currently fielded low-velocity airdrop system.

The need for airborne forces to land relatively close to their objective areas also has important implications for the Air Force because Air Force transport aircraft (C-130, C-5, and C-17) are used to deploy Army airborne forces. Objectives important enough to be targets for airborne assault will likely be defended against both air and ground attack. The closer that Air Force transport aircraft must operate to the objective area, the greater this threat will probably be. Therefore, the current limitations of today's Army airborne forces affect the employment—and threat exposure—of Air Force transport aircraft. Importantly, it is unlikely that Air Force transport aircraft will be risked in areas where unsuppressed radar-guided surface-to-air missiles (SAMs) could threaten the aircraft.

CHAPTER THREE

Threats to Today's Airborne Forces

Historically, airborne operations have had a fairly high degree of risk, particularly if the operation takes place against a reasonably competent opponent. Most large-scale airborne operations in WWII resulted in high numbers of casualties. For example, during D-Day airborne operations on June 6, 1944, the 82nd and 101st Airborne Divisions parachuted or landed by glider a total of roughly 13,000 personnel; it was later calculated that the 82nd suffered 1,259 casualties (killed, wounded, or missing) and that the 101st suffered an additional 1,240 during the first 24 hours of the operation, a combined casualty rate of 19 percent.[1] Three months after D-Day, the combined British-American armored/airborne assault into Holland, Operation Market Garden, also led to heavy losses for the airborne forces, particularly the British 1st Airborne Division, which suffered more than 7,000 casualties out of some 10,000 personnel participating in the operation.[2]

Some airborne operations in the WWII era produced very important results, but often at great cost. One reason for the heavy casualties was the fact that the foot-mobile units of that era typically had to drop fairly close to their objective areas, primarily locations that were important to the enemy and therefore heavily defended.

Since the end of WWII the largest U.S. airborne operations have been brigade-size. Fortunately, these operations have not experienced

---

[1] Gordon A. Harrison, *Cross Channel Attack*, Washington, D.C.: U.S. Army Center of Military History, 1951, p. 284.

[2] Mark Fielder, "The Battle of Arnhem (Operation Market Garden)," BBC History, February 2, 2011; Cornelius Ryan, *A Bridge Too Far*, New York: Simon and Schuster, 1974.

casualties of WWII magnitude, nor have the post-WWII operations faced opponents of the quality of the German Wehrmacht. That said, even battalion-sized airborne operations have been relatively rare in the past several decades.

Today and in the foreseeable future, several classes of threats can challenge airborne operations. Those threats will be reviewed below, including the possible effect that each type of threat has on the planning and conduct of future airborne operations.

Improved Air Defenses

Ground-based air defenses have been a challenge to airborne operations since the 1940s. Occasionally (as in Crete in May 1941, for example) air defenses have inflicted very heavy casualties on airborne forces, in large part due to the need for foot-mobile airborne units to land relatively close to their objectives. In addition to actually inflicting casualties, air defenses can result in the cancellation of airborne operations simply due to the fact that a threat exists, as was the case in Operation Iraqi Freedom in 2003.[3]

There are two general classes of air defenses that influence airborne operations: medium-/high-altitude defenses and low-altitude defenses.

Medium-/High-Altitude Defenses

Medium-/high-altitude defenses generally apply to aircraft flying above 15,000 feet. From WWII until the early 1960s, it was heavy anti-aircraft guns that challenged aircraft at those altitudes; today, radar-guided SAMs epitomize this class of threat. Some medium-/high-altitude SAMs are relatively short-range in terms of the horizontal distance from the launcher to the target aircraft. The very high-quality Russian-built 9K330 Tor system (North Atlantic Treaty Orga-

---

[3] Marc DeVore, *The Airborne Illusion: Institutions and the Evolution of Postwar Airborne Forces*, working paper, Cambridge, Mass.: Massachusetts Institute of Technology, June 2004, pp. 18, 28–31.

nization [NATO] designation SA-15; see Figure 3.1) is an example of this class of radar-guided SAM. With a maximum altitude of more than 20,000 feet and a range of roughly 12 km from launcher to target, the SA-15 is highly resistant to electronic countermeasures, can track multiple targets simultaneously, and is able to fire while on the move. Individual Tor launcher vehicles can also operate independently if necessary. The SA-15 missile system can turn its radar off, displace to a different firing position, and be ready to fire from a new location in under ten minutes. Consequently, the SA-15 is a system that poses a significant challenge to reconnaissance and strike platforms supporting airborne forces.

Other larger, radar-guided SAMs have much longer range. The Russian-produced S-300/400-series (NATO designation SA-10/20/21) SAMs are much larger than the SA-15 and have ranges of up to 400 km, depending on the specific model of missile that is being fired. These large systems operate in multivehicle battery-sized units and require just a few minutes to emplace and prepare to fire, and they have a larger support "tail" than smaller systems, such as the SA-15.

Figure 3.1
SA-15

SOURCE: Vitaly V. Kuzmin, CC BY-SA 3.0.
RAND *RR309-3.1*

Their great range, accuracy, and resistance to electronic countermeasures, however, make them a very serious threat to the large, relatively slow transport planes that carry an airborne force.

The great range of modern SAMs, such as the SA-10/20, means that the threat can now come from a much larger area. Whereas a 1960s- or 1970s-era SA-6 (then a state-of-the-art Soviet air defense system) had a maximum range of about 25 km and could thus cover an area of roughly 1,900 square km, an SA-20 with a range of 200 km is able to threaten aircraft in an area more than 65 times that size: over 125,000 sq km. From a planning perspective, this means that a much, much larger area must be "suppressed" today than in previous years. It also means that there is a far larger area to saturate with accompanying intelligence, surveillance, and reconnaissance (ISR) assets or supporting on-call fixed-wing attack platforms. The Air Force and Navy could expect nonstealthy fighters to take serious losses from unsuppressed modern radar-guided SAMs, such as the S-300/400. Transport aircraft would, of course, be at far greater risk than fighter aircraft if this class of threat were present.

The disadvantage of radar-guided SAMs is that they must "emit" to acquire and engage targets. This is one of the main differences between the medium-/high-altitude and low-altitude threats. Once a radar turns on, it can be located. Since the Vietnam War, the Air Force and Navy have devoted considerable effort to the location, suppression, or elimination of this kind of threat by fielding airborne tactical jamming systems, radio countermeasure sets, and anti-radiation missile systems, such as the AGM-88 high-speed anti-radiation missile.

The challenge of suppressing air defenses should not, however, be underestimated. During Operation Allied Force in Serbia-Kosovo in 1999 the NATO air forces had great difficulty in locating the Serb air defenses, including emitting radar-guided SAMs. Clever radar management and frequent moves of SAM units on the part of the Serbs reduced the effectiveness of NATO's suppression of enemy air defense systems operation. The presence of unsuppressed air defenses in Kosovo also helped convince U.S. and NATO senior decisionmakers that the use of the attack helicopters of Task Force Hawk would be too risky, and that air assault operations into Kosovo were not a viable option unless the air defenses could be significantly reduced.[4]

In the future, depending on the resources and level of expertise of the opponent, suppressing or eliminating the medium/high-altitude SAM threat could be a time-consuming and resource-intensive effort that is reactionary in nature. The perceived success, or lack thereof, of the suppression of enemy air defense systems operation against this class of threat would be of profound importance to senior decision-makers who are considering an airborne operation. In the recent past, including in Iraq in 2003, airborne operations were canceled due to a lack of confidence in the level of suppression achieved against the enemy's air defenses. It is very unlikely that senior U.S. commanders would be willing to take the risk of exposing relatively slow transport planes in an area where modern, radar-guided SAMs may be operating.

While this threat is potentially serious, it should be noted that relatively few countries have modern medium-/high-altitude defenses. An important factor of modern medium-/high-altitude SAMs is their high cost. Most nations cannot afford many, if any, of these systems. The unit cost of an SA-15 is roughly $25 million per launcher vehicle, and a battery of SA-10 SAMs has been estimated to cost roughly $100 million, including a basic load of missiles.[5] So, while the most advanced SAM systems would pose a severe challenge to airborne operations, they are likely to be absent or few in number in most scenarios. According to one source, there are currently 20 countries (including Venezuela, Syria, and North Korea, among others) that either have the S-300 (SA-10) SAM or are planning to acquire the system.[6]

---

[4]   Bruce R. Nardulli, Walter L. Perry, Bruce R. Pirnie, John Gordon, and John G. McGinn, *Disjointed War: Military Operations in Kosovo, 1999*, Santa Monica, Calif.: RAND Corporation, MR-1406-A, 2002, pp. 28–30, 94–95.

[5]   "News Archives: Chinese Missile Defenses," *Missile Threat*, undated.

[6]   International Institute for Strategic Studies, *The Military Balance, 2012*, London, March 2012.

Low-Altitude Defenses

The other class of air defenses that can affect airborne operations is the low-altitude threat. The nature of this challenge is very different compared to the medium/high-altitude threat. Whereas weapons such as the SA-15 or SA-10/20 must turn on their radars to acquire, track, and engage high-flying aircraft, that is not the norm for low-altitude defenses. In fact, this class of air defense threat is really the primary concern for future airborne planners because, should modern radar-guided SAMs be within range of the possible drop zones or the flight paths of transport planes, it is unlikely that senior joint commanders will authorize an airborne operation.

Low-altitude air defenses are mostly of two types: anti-aircraft artillery (AAA) consisting of 14.5- to 57-mm guns (see Figure 3.2), and man-portable air defense systems (MANPADS). The former are single-, dual-, triple-, or quadruple-barreled guns that are mounted either on the ground or on a vehicle. MANPADS are generally small, shoulder-fired missiles that are usually infrared-guided, though the latest generation of MANPADS are true image seekers. MANPADS can easily

Figure 3.2
S-60 57-mm Anti-Aircraft Gun

SOURCE: Wikimedia Commons user Bukvoed, CC BY 2.5.
RAND RR309-3.2

be carried by one person but are occasionally mounted on light vehicles. Compared with the large systems associated with medium-/high-altitude air defenses, low-altitude weapons are much smaller, far easier to conceal, available in much larger numbers due to their relatively low cost, and they are usually optically directed. The last characteristic is very important, since low-altitude weapons normally do not have to turn on their radar to acquire or engage an aircraft. Their passive nature makes them very difficult to locate prior to being fired. Compared with radar-guided SAMs, however, low-altitude defenses generally have much shorter ranges. Most MANPADS, for example, have ranges of less than 5 km, and gun ranges are usually less than that.

AAA is generally immune to countermeasures. Once a gun fires at an aircraft, there is little the aircraft can do; the rounds will either hit or miss. In terms of protecting the aircraft against AAA hits, on some aircraft types, certain critical components can be hardened against hits up to 23 mm. Once an AAA system reaches roughly 30 mm, however, the amount of armor required to protect against a projectile of that size is prohibitive in terms of aircraft weight.[7] The trend in modern AAA is toward guns of 30 mm or larger. Most AAA is limited to a vertical range of less than 10,000 ft (3,048 m).

It should also be noted that while most AAA systems are in the 14.5- to 57-mm range, there are still some older, large-caliber guns in use around the world. These include 85- and 100-mm Soviet-era and Chinese weapons. Very similar to the heavy anti-aircraft guns of WWII, these weapons are gradually being phased out but may still be encountered in some countries.

MANPADS, which are normally infrared-guided, can be "spoofed" by various types of countermeasures, such as flares or infrared jamming devices, but the susceptibility to countermeasures varies greatly depending on the type of target aircraft and the model of the MANPADS. Older MANPADS, such as the 1960s-era SA-7 missile system that has an uncooled lead sulfide seeker and is capable of rear-

---

[7] John Gordon, Peter A. Wilson, Jon Grossman, Dan Deamon, Mark Edwards, Darryl Lenhardt, Daniel M. Norton, and William Sollfrey, *Assessment of Navy Heavy-Lift Aircraft Options*, Santa Monica, Calif.: RAND Corporation, DB-472-NAVY, 2005, p. 64.

aspect shots only, are relatively easy to decoy away from an aircraft with today's countermeasures. More modern, sophisticated weapons such as the SA-18 Igla (Figure 3.3) are much more resistant to countermeasures due to their more advanced seekers, enhanced lethality and improved performance, and terminal programming guidance.[8] Typically, MANPADS have a maximum altitude of 15,000 ft (4,572 m) or less. Unfortunately, MANPADS are widely available on the global arms market and range from older models, now easy to counter, to more modern systems that are a challenge for countermeasures. After the collapse of the Qaddafi regime in Libya, large numbers of MANPADS ended up on the global weapons black market, meaning that these weapons are also likely to be in the hands of nonstate actors.

One challenge of this class of air defenses for airborne operations is that there is very little reaction time to launch countermeasures, such as flares, or conduct evasive maneuvers once the missile is in the air. Another is that they can be ubiquitous and are very difficult to locate before they are actually fired. In Kosovo in 1999, for example,

Figure 3.3
SA-18

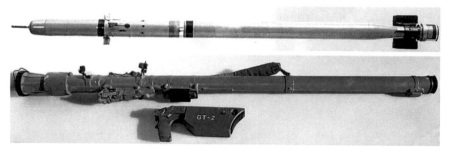

SOURCE: : U.S. Department of Defense photo.
RAND RR309-3.3

---

[8]   Early-generation MANPADS, such as the SA-7, have a lead sulfide seeker head that is sensitive to electromagnetic energy between 1 and 2.5 micrometers. This range corresponds to shorter infrared wavelengths wherein only "hot" objects, such as flares, emit radiation. Later-generation MANPADS, such as the SA-18, have seekers coated with lead sulfide and indium antimonide, which is sensitive to electromagnetic energy out to 5 micrometers. This allows the MANPADS missile to distinguish between hot flares and the heat signature of the aircraft.

Serb forces had hundreds of anti-aircraft guns and MANPADS—far more than the number of radar-guided SA-6s operating inside Kosovo. The presence of these weapons forced NATO aircraft to operate above 10,000 feet and heavily influenced the decision not to employ attack helicopters there.

In response to the MANPADS threat the Air Force now mounts automated countermeasures on transport aircraft. For example, the Large Aircraft Infrared Countermeasures System employs lasers against MANPADS, along with older systems that automatically eject flares when an approaching missile is detected.

Another element of low-altitude defenses is small arms—the rifles, machine guns, and man-portable antitank weapons that enemy troops carry. Although they are less effective against aircraft flying more than a few hundred feet above the ground, the sheer numbers of these weapons can pose a considerable threat to low-altitude aircraft, especially low, slow-flying aircraft approaching a landing zone, drop zone, or airfield. In Vietnam, rocket-propelled grenades (RPGs) shot down several hundred U.S. helicopters, for example.[9] In Iraq in 2003, an attack by the Army's 11th Attack Helicopter Regiment was confronted with sudden, massed small-arms and rocket-propelled grenade (RPG) fire from all directions. Every helicopter was hit by enemy fire; one was lost and many were seriously damaged.[10] The losses suffered in that mission prompted considerable caution on the part of Army senior commanders in Iraq, who for the rest of the operation displayed a reluctance to employ low-altitude aircraft when there was the possibility of a similar threat.[11]

Transport planes carrying Army airborne units are different from helicopters, particularly in terms of speed and ability to fly at higher altitudes. However, transport aircraft can still be threatened by low-altitude air defenses, particularly during their approach to and departure from drop zones, when the aircraft are below a few thousand feet

---

[9]   Carlo Kopp, "Are Helicopters Vulnerable?" *Australian Aviation*, March 2005.

[10]   Greg Grant, "A Little Bird for the Army?" *DoD Buzz*, October 23, 2008.

[11]   John Gordon, David Johnson, and Peter A. Wilson, "Air Mechanization: An Expensive and Fragile Concept," *Military Review*, January–February 2007.

altitude and at speeds of less than 200 knots. Additionally, if those same aircraft are at higher altitudes during what could be a lengthy flight over enemy territory as they are headed toward the drop zones, or back home after a drop, they may also expose themselves to medium-/high-altitude air defenses.

Implications of These Threats

The important point about air defenses is that they present a multi-faceted challenge to airborne operations. The low- and medium/high-altitude threats are very different in nature and require different types of countermeasures. The longer range of today's medium-/high-altitude defenses means that a much greater area will have to be subject to suppression in order to protect the transport aircraft. A much larger area also means larger numbers of dedicated and dispersed on-call fixed-wing attack platforms will be needed to support the airborne package.

Airborne forces are much more viable against a poorly armed opponent (whether state or nonstate) that has only limited air defense capability than against well-armed enemies that can threaten aircraft with a mix of high- and low-altitude air defenses. Importantly, when unsuppressed, modern radar-guided SAMs are suspected of being in the vicinity of a proposed airborne operation, it is likely that senior decisionmakers will not be willing to take the risks associated with such an operation.

The farther airborne forces can drop from the worst of the enemy's defenses, the more both of these risks can be managed, the greater the likelihood that decisionmakers will be willing to take those risks, and the greater the probability of mission success—as long as the airborne force is able to conduct its mission from drop zones farther from the objective than is the norm today.

Long-Range Fires Directed Against Drop Zones and Airheads

Many airborne operations are initiated with a parachute operation to seize a lodgment for follow-on forces to arrive. Often this includes the seizure of an airfield for follow-on forces to fly into. The 1983 (Grenada) and 2003 (northern Iraq) airborne operations both employed this technique. Today's 82nd Airborne Division often plans operations in that manner: Parachute forces seize an airhead and secure it for the arrival of other, heavier follow-on forces. Indeed, for years the XVIII Airborne Corps' operational concepts have included the possibility of small numbers of M-1 Abrams main battle tanks and M-2 Bradley infantry fighting vehicles being air-landed into airheads that were seized by parachute units, as took place in northern Iraq.

One of the growing challenges to today's airborne operations is the increasing range and precision of long-range indirect fire systems that can threaten the vulnerable aircraft operating in and out of an airhead. As was the case with today's SAMs, the trend in surface-to-surface indirect fire systems is toward longer range. For example, in the 1960s–1970s, the 40-barrel, 122-mm BM-21 was the standard multiple-rocket launcher (MRL) of the Soviet army. The BM-21 had a maximum range of 20 km. Today's MRLs have much longer ranges.

The Russian BM-30 Smerch exemplifies the type of truck-mounted MRLs that are available today. Today the 12-barrel Smerch has a maximum range of roughly 90 km, which is being increased to 120 km. Its 300-mm rockets can carry a mix of warheads, including submunitions. Submunitions delivered from long range are a particularly serious threat to an airfield, since a single battery of BM-30 launchers could saturate a runway and aircraft parking area in a few seconds of firing. Transport planes on the ground are highly vulnerable. Even if no aircraft were hit, if the runway were littered with small fragments from exploded submunitions, air operations would be shut down until the runways were cleared. Even with a dedicated runway repair team, this could take several hours.[12] On top of that, there is a potential problem of unexploded ordnance, which could require explosive ordnance disposal expertise to deal with, thus causing more delays.

---

[12] While Air Force repair squadrons are capable of fixing damaged runways in a few hours, the Air Force does not typically plan to deploy those units into just-seized, expeditionary airfields.

This could result in the need for more engineering assets to be deployed into an airhead to keep the runways in operation.

Even middle-tier militaries are building this type of weapon. The Iranian-built Fajr-5 is a 333-mm MRL mounted on a Mercedes truck. The rocket has a maximum range of roughly 75 km and has several warhead options. Like other MRLs, the weapon can be rapidly emplaced, fire, and then quickly depart from the firing position. If the target is an airfield, precise accuracy is not necessary. The arrival of several dozen rockets within a span of a few seconds, scattering submunitions across a runway and aircraft parking ramps, would shut down airfield operations, possibly for several hours. If transport planes were destroyed on the ground by such a barrage, decisionmakers' willingness to continue to fly aircraft into the airhead would almost certainly be diminished, if not ended, until the threat was dealt with. If these weapons were fired on an active drop zone, personnel would be at grave risk. It should be noted that the latest version of Fajr-5 has a considerably longer range than the original model.

Depending on the type and model of the rocket launcher, the number of rockets that would have to be fired to have a good chance of destroying aircraft on the ground at an airfield would vary considerably. For example, the Chinese-built WS-2 MRL is claimed to have an accuracy of roughly 200 meters at a range of roughly 150 km. That system, like the Russian Smerch, would be a formidable threat, able to target specific areas on an air base. The Iranian Fajr-5 (Figure 3.4), however, is a much less accurate system that would have to fire more munitions to have a high probability of damaging key facilities at an air base or destroying aircraft on the ground.[13]

Recent experience in Iraq and Afghanistan demonstrates that rockets can be emplaced in makeshift launchers and fired on a timer or by remote control by irregular forces. Often, irregular forces will use weapons of this type to harass airfields. Similar weapons were used as early as the Vietnam War in the 1960s, when communist forces fired rockets into U.S. air bases in South Vietnam.

---

[13] "Weishi (WS-1/-2) Multiple Launch Rocket Systems," web page, last updated December 31, 2008.

Figure 3.4
Fajr-5

SOURCE: Iranian Ministry of Defense photo.
RAND *RR309-3.4*

Another important aspect of this class of threat is the fact that, compared with almost all U.S. Army organic indirect fire systems, today's foreign-built MRLs have much greater range. The Army's Multiple Launch Rocket System has a maximum range of 84 km with a guided rocket. In contrast, the Chinese WS-2 400-mm rocket launcher has a range of over 200 km. Army 155-mm cannon artillery (either the towed M198 or the self-propelled M109A6 Paladin) is limited to 30 km, even when firing rocket-assisted projectiles. Only the Army Tactical Missile System had ranges greater than modern foreign MRL systems, but it is no longer produced.

Until and unless the Army can develop a new, strategically mobile longer-range indirect fire system, its airborne forces will remain at a disadvantage against long-range MRLs. In the interim, the Army will be largely dependent on fires provided by supporting fixed-wing aircraft from the Air Force and Navy to counter the MRL threat. Army Firefinder radars deployed into an airhead may be able to locate enemy MRLs firing against the lodgment area (depending on the range of the hostile system), and those data would be passed to nearby fighters or bombers. This technique would, however, require that supporting aircraft from the other services be constantly available to respond to a sudden MRL barrage and would still allow the enemy to get an initial barrage off against the target airfield. In some circumstances, attack helicopters (organic to the Army or Marine Corps flying from ships in the operational area or from nearby land-based locations) might be available to attack enemy MRLs, depending on how far the airborne lodgment area is from helicopter operating areas. However, if firing positions are tens of kilometers from the target, firing elements could move before aircraft were able to interdict them.

Joint fires would still be needed, but as was the case with enemy air defenses, the long-range indirect fire threat could also be mitigated to some extent by selecting drop zones and airheads that are more distant from the objective than is the case today.

Army airborne forces may also need an air-deployable defensive system to intercept incoming rockets. This could be, for example, a deployable version of the "Iron Dome" defensive system that is currently in use in Israel to defend cities and towns from incoming rockets. The original "Iron Dome" system produced by Israel with Raytheon was intended to protect population centers with little need for tactical or strategic mobility. A new, more deployable system called "Battle Dome" that might be more appropriate for airborne units that have to deploy into lodgment areas is being developed by Raytheon.[14] Other defensive systems, including lasers and guns, might also be appropriate for countering this type of threat.

The Ground Threat to Airborne Operations

In today's environment there are generally three types of opponents the U.S. military could face: *irregular*, *hybrid*, and *state militaries*. Depend-

---

[14] Discussions with Raytheon representatives, RAND Corporation, Santa Monica, February 2013.

ing on the type of the opponent, the ground threat to U.S. airborne forces could vary greatly.[15]

Irregular opponents are not likely to have many, if any, traditional weapons systems such as armored vehicles, heavy artillery, or aircraft. That said, hostile irregular forces could be well equipped with small arms (including sniper weapons), improvised explosive devices (IEDs), light antitank weapons of the RPG class, and light indirect fire systems, such as mortars.[16] These systems could pose a considerable threat to U.S. light infantry forces and, to some extent, mechanized units (e.g., the IED threat). Depending on the skill of the opponent, this class of foe could still inflict considerable casualties on a U.S. force, particularly if the Americans lacked adequate protection.

Hybrid opponents have the ability to fight simultaneously as irregular and conventional forces. Therefore, the types of threats just listed for irregular forces also apply to hybrid opponents. Additionally, hybrid forces can employ at least some of the techniques and systems available to the conventional forces of an enemy nation. This could include some air defenses beyond just small arms and various versions of MANPADS, as well as more advanced antitank guided missiles, combat vehicles, and heavier indirect fire systems.

At the highest end of the spectrum would be the forces of an opposing nation. Ironically, these forces would be the easiest for U.S. ISR assets to locate, since the enemy personnel wear uniforms, they operate large and identifiable military equipment, and they are employed in recognizable units and doctrinal patterns. That said, the forces of an enemy nation probably have the greatest level of military capability with which to oppose U.S. airborne forces, particularly air defense and indirect-fire weapons.

It has already been pointed out that one of the ways to at least partially mitigate longer-ranged enemy air defenses and long-range indirect fire threats to airheads would be to seek greater offset from the ultimate objective area: dropping the airborne force farther from what

---

[15] Frank G. Hoffman, "Hybrid Warfare and Challenges," *Joint Force Quarterly*, No. 52, 2009.

[16] They could also have MANPADS and AAA weapons at their disposal.

the enemy cares about and will probably defend. If this is possible, the airborne force would then have to maneuver from the drop zones and airheads toward the objective, possibly a considerable distance. Some or all of these ground threats listed could be encountered during the advance toward the objective area. Therefore, future airborne forces seeking to mitigate threats to their insertion will need both appropriate capabilities to maneuver greater distances than their generally foot-mobile predecessors and an appropriate level of protection and firepower to overcome opposition between their drop zones and airheads and the objective.

Of particular concern to a future light armored airborne infantry force could be the anti-armor weapons available to an opponent. These could include RPGs of various types, antitank guided missiles, mines, and guns mounted on armored vehicles. In general terms, a light armored airborne force would try to either maneuver around significant areas of opposition as it headed toward its objective or overwhelm the enemy as quickly as possible with a combination of organic and joint fires. It is likely that an airborne unit dropped in the opponent's "rear area" would encounter relatively lightly armed logistics units, air base personnel, and other support troops who would be poorly prepared to deal with the sudden appearance of light armor and motorized infantry. That said, the relatively light armored vehicles associated with an airdroppable future airborne force would require the use of appropriate tactics when encountering opposition that included anti-armor weapons.

CHAPTER FOUR

A Proposed Airborne Light Armored Infantry Force

Overview of the Concept

To mitigate many of the threats described in Chapter Three, the concept developed by the RAND research team focuses on introducing light armored vehicles into today's airborne force. Organic light armor has been lacking in Army airborne units since the retirement of the Sheridan.

The concept described here involves more than just reintroducing light tanks (or an equivalent vehicle) into the airborne force. Even when the Sheridans were present in the 82nd Airborne Division, the majority of the division still maneuvered at the pace of a walking infantryman. Today, with the need to seize lodgments outside the range of most of the surface-to-surface fires threats, as well as the bulk of the enemy's low-altitude air defenses, forces require a generally higher level of mobility, including the ability to fight on the move with some degree of armor protection. While portions of the infantry of today's airborne forces have mobility by means of airdroppable High-Mobility Multipurpose Wheeled Vehicles (HMMWVs), the lack of protection and the inability to conduct mounted, mobile combat in those vehicles leave a considerable capability gap in today's airborne infantry units. Therefore, the concept developed by RAND includes a more robust role for light armored vehicles in the airborne force.

It should be noted that some other airborne forces, particularly Russia's and China's, have organic light armor. Russia has employed light armor in its airborne units since the 1960s, when the ASU-57 self-propelled antitank gun became part of its parachute units. Later, it added specialized airdroppable fighting vehicles, such as the BMD infantry fighting vehicle. The Chinese took a similar path, and today's airborne units in the People's Liberation Army include light armored vehicles that are parachute-droppable. Today, the Russian Air Force has enough transport aircraft to airdrop one 5,000-personnel, two-regiment airborne division, including its fighting vehicles, in roughly three sorties.[1]

Key elements of the concept proposed in this report include the following:

- A suite of light armored vehicles for a variety of roles, such as light tanks, armored personnel carriers, mortar carriers, command-and-control vehicles, ambulances, reconnaissance vehicles, and anti-armor vehicles. The vehicles would be able to fight on the move and would provide limited armor protection for crews and passengers.

- All or most of these vehicles would be capable of parachute drop from existing Air Force transport planes. Ideally, as many of the vehicles as possible would be capable of deploying by C-130 (including by airdrop), taking advantage of the large number of C-130s that are qualified for parachute operations.

- Airborne forces equipped with these vehicles would be able to drop outside the bulk of the enemy's air defenses and surface-to-surface fires, employing the mobility, firepower, and protection of the vehicles to maneuver toward their objectives.

- The actual distance from the drop zone(s) to the objective would vary considerably, depending on the threat, the terrain, and the degree of urgency associated with accomplishing the specific mission. Distances of 30–100 km or more would be feasible given the increased mobility of these future airborne units.

- The purpose of the initial airdrop could be either to deliver the airborne unit, which would quickly "abandon" the drop zone and

---

[1] Rod Thornton, *Organizational Changes in the Russian Airborne Forces: The Lessons of the Georgian Conflict*, Carlisle, Pa.: Strategic Studies Institute, U.S. Army War College, December 2011.

start moving toward its objective or to prepare for the arrival of additional forces in the immediate vicinity in preparation for the move toward the ultimate objective.

- The maneuvering airborne units equipped in this manner would rely heavily on joint fires (e.g., close air support) and ISR systems, such as unmanned aerial systems for situational awareness.

- As the airborne light armored infantry force maneuvers toward its objective from drop zones that would be farther from the objective than has typically been the case in the past, it would rely on aerial resupply from Air Force transport planes, delivered accurately from medium altitude using the Joint Precision Airdrop System (JPADS). Today, JPADS has a maximum capacity of 10,000 lbs for a fully rigged pallet, and a C-17 can drop up to 110,000 lbs of JPADS pallets. Accuracy is measured in tens of meters from the intended delivery point. Because drop accuracy with JPADS is much less sensitive to aircraft location, this technique would allow Air Force transport planes to remain well above AAA and MANPADS range while carrying out airdrop missions.

- An airhead would almost certainly still be required for the arrival of air-landing follow-on forces and portions of the required resupply, but due to the protected mobility of the airborne units, the airfield(s) selected for seizure could be farther from the objective than is the norm in today's airborne operations.

Another consideration is how much of today's airborne force should be converted to this configuration. If it appears that the basic concept of increasing the level of protection, mobility, and firepower for some portion of the airborne force via the introduction of airdroppable light armor in combination with other, unarmored, vehicles is a sound move for the Army, one of the next steps would be to determine how much of the force should be converted to this configuration. Options include entire airborne brigades (although the amount of air lift that would be required to move an entire brigade would be very large), or one or more battalion-sized units in the 82nd Airborne Division. Additionally, portions of the 173rd Airborne Brigade and the 4th Brigade of the 25th Infantry Division could be enhanced in this manner.

Light Armored Vehicle Options

Given the near-term focus of this project, and therefore the need to consider extant equipment, the Light Armored Vehicle, second generation (LAV-II) family of vehicles appeared to be the most promising candidate platform for enabling an airborne force to conduct mounted maneuver. The LAV-II series offers advantages in two broad categories. First, it is suitable as an airborne combat vehicle: It can be airdropped from the C-17 Globemaster III, and some variants of the LAV-II can be airdropped from the C-130 following minor modifications as was done in 1991 when the 82nd Airborne Division tested the Marine Corps' LAV-II. For a vehicle in its weight class, it has good all-around protection (14.5 mm on the frontal arc, 7.62 mm all around), firepower, and mobility. Second, the LAV-II is also available from a U.S.-aligned vendor; it is currently fielded in the U.S. Marine Corps (as LAV-25A2 and variants) and would be available for use in short order, rather than requiring a full, new development effort.

The research team also considered the Stryker-series vehicles currently operated by the U.S. Army. If airdrop is an important consideration for the light armor of airborne forces, Stryker (a derivative of the LAV-III) has significant limitations, described later in this chapter.

LAV-II

There are numerous variants of the LAV-II, which is a potential advantage for offering a range of capabilities to airborne commanders. In Marine Corps light armored reconnaissance battalions, the primary vehicle employed is the LAV-25, a scout vehicle armed with a turreted 25-mm Bushmaster cannon that can carry three to four dismounts in the rear of the vehicle. The Marine Corps also employs a command variant, a mortar variant with an 81-mm mortar, a logistical variant without a turret, an antitank variant armed with the TOW-II, and an electronic warfare variant. The Marine Corps has also tested an air defense version armed with a 25-mm minigun and Stinger missiles, as well as a version armed with a 120-mm mortar system called Dragon Fire. In addition to the Marine Corps' LAV-IIs, the Saudi Arabian National Guard has purchased a variety of LAV-IIs, including an assault gun variant with a turreted 90-mm cannon and a mortar vehicle with a turreted 120-mm mortar.

General Dynamics Land Systems has continued to develop and update the LAV-II. There is a technology demonstrator called LAV-IIH that is an armored personnel carrier variant with an extended hull that can carry a full squad of nine dismounts plus a two-person crew; it has countermine and IED upgrades, and a dual-v hull (DVH) version has been offered. An improved engine with substantially better fuel efficiency is also available. It should be noted that the Marine Corps did not purchase an armored personnel carrier (APC) version of the LAV-II because its vehicles are organized in light armored reconnaissance battalions that have only a small number of personnel ("scouts") that dismount from the LAV-25s.

The LAV-II's characteristics should permit three of the vehicles to be airdropped from C-17. Because it is narrow enough to fit side by side in the C-17, up to five may be transported for air-landing in a single airplane (or seven in the C-5). The LAV-II can be transported by the C-130, and some versions can be airdropped from that aircraft; it was tested for both parachute drop and Low Altitude Parachute Extraction System delivery in the late 1980s and early 1990s. Indeed, the 82nd Airborne Division was experimenting with 14 LAV-25s provided by the Marine Corps when Iraq invaded Kuwait in 1990. The 82nd deployed to Saudi Arabia and used the vehicles in combat.[2] Some variants may require a waiver for height issues to be dropped by parachute (low-velocity aerial delivery, or LVAD). In 1990–1991, the Marine Corps' LAV-25s loaned to the 82nd Airborne Division had their turrets lowered by four inches to meet the height limitations of the C-130's cargo bay.

Stryker

The Stryker family of vehicles currently in Army service could be viewed as an alternative to the LAV-II but pose a number of significant

---

[2] Information provided by representatives of the Dominant Maneuver Division, Office of the Deputy Chief of Staff, G-8, Force Development Headquarters, Department of the Army, the Pentagon, January 2013.

challenges for use with airborne forces. The Stryker's chief advantage is that it is already in the Army inventory. However, its disadvantages argue against its inclusion, except possibly as a supplemental vehicle in the air-landing echelon of an operation. For instance, the Stryker requires more aircraft to transport or airdrop the same number of vehicles, currently no cargo parachute system that can drop Stryker is being fielded or in development, and new variants of Stryker would need to be procured to obtain appropriate organic direct firepower for the forced entry mission.

The M1126 Stryker infantry carrier vehicle (ICV) is larger and much heavier (over 38,000 lbs) than the LAV-II (roughly 30,000 lbs). This would restrict its use for airdrop to two per C-17, for reasons of both mass and volume. When rigged for airdrop, the Stryker would exceed the current maximum weight for LVAD that is limited to 42,000 lbs. A 50,000- to 60,000-lb heavy LVAD capability would need to be tested and fielded before the various Stryker models could be dropped by parachute. For example, when a single mobile gun system (MGS) version of Stryker was experimentally parachuted in 2004 from a C-17, waivers had to be obtained, since the total weight (vehicle, platform, rigging, and parachutes) was over 52,000 lbs. While the 60,000-lb LVAD was partially developed in the 1990s (and the prototype was used in a quick-response effort to test the Stryker MGS variant), no current requirement exists in support of LVAD beyond 42,000 lbs, and completing testing and fielding it would take a number of years.[3] Discussions with the developers of Army parachute technology at Natick Laboratory indicated that there is no formal Army requirement to increase the current LVAD limit beyond 42,000 lbs today.

The Stryker is also much less suitable for use with the C-130 than is LAV-II. No Stryker variant can be dropped from the C-130, and the

---

[3] A 60,000-lb-capacity LVAD system was developed by the Army in 1994 but never fielded. In a 2004 test of the ability of C-17 to airdrop the Stryker MGS variant, it was used but deemed "unsatisfactory for use," in part due to damage to the aircraft and platform. The Army has not articulated a requirement for completing the fielding of this capability, and no work is being done on aerial delivery of loads greater than 42,000 lbs (Air Force Flight Test Center, *Engineering Feasibility Assessment of C-17A Aerial Delivery of the U.S. Army Stryker Mobile Gun System Vehicle*, AFFTC-TR-04-38, November 2004, pp. 2–3, 6).

latest DVH versions cannot even be transported by the C-130.[4] The ICV is not regularly carried in C-130s, as it must be partially disassembled for transport and its weight significantly affects the C-130's range and landing profile.

Finally, for the forcible entry mission, it is desirable that an airborne light armored infantry force have the ability to fight mounted during the phase of the operation when the force is maneuvering from drop zone to objective. The Stryker BCT (SBCT) organization emphasizes protected mobility for a force that is intended to fight dismounted. While a limited number of 105-mm cannon–armed MGSs are available, the bulk of the SBCT's direct firepower comes from dismounted infantry and remote weapon stations armed with heavy machine guns and grenade launchers. Part of the value of having Stryker vehicles already in the inventory would be offset if new variants with heavier weapons needed to be procured.[5]

By contrast, the LAV-25 is readily available and has desirable characteristics for the screening/reconnaissance as well as infantry fire support roles for an airborne force. In that regard, it should be noted that the Army did not procure versions of the Stryker with 25-mm guns, which are available in the LAV-25A2.

HMMWV

Although no Army airborne units currently have the LAV-II, the 82nd Airborne Division in Fort Bragg and the 173rd Airborne Brigade in Vicenza, Italy, are each equipped with airdroppable HMMWVs. Discussions with representatives from the 82nd revealed that the Division can field a total of roughly 1,200 HMMWVs, 22 of which are kept rigged for airdrop on 16-foot platforms. Several 105-mm howitzers are also kept rigged for airdrop on similar platforms, to be attached to the vehicles after landing.

---

[4] See General Dynamics Land Systems, undated(c).

[5] The Army currently has no Strykers armed with 25-mm guns. Assuming a light armor airborne force would require a fight-on-the-move capability, the Army would have to procure new versions of Stryker to gain it.

Given the fact that the 82nd Airborne Division consists of eight infantry battalions of 600–800 personnel each, the current number of HMMWVs clearly is not enough to transport a large percentage of infantry. Discussions with the XVIII Airborne Corps also revealed that current plans actually *reduce* the present number of HMMWVs in the 82nd Airborne, thus making the infantry units even less mobile. Representatives made the following point about the division's infantry: "They walk to the objective."[6]

Importantly, while an increased number of HMMWVs or similar vehicles would improve the tactical mobility of airborne infantry units, those vehicles have very limited protection and restricted off-road mobility, and they cannot fight on the move. That said, one of the options to enhance future airborne forces could be a mix of light armored vehicles for a portion of the force, with a wider use of unarmored vehicles, such as the HMMWV, for the infantry and other elements that currently have few or no vehicles today.

Airlift Requirements

This section examines the airlift requirements for the four main airborne force packages that were identified in this study: (1) Stryker-based brigade, (2) LAV-based brigade, (3) Stryker-based battalion, and (4) LAV-based battalion. These airborne force packages were built by starting with the current Stryker battalion/brigade Table of Organization and Equipment (TOE) and modifying them to suit the airborne light armor force concept. In all cases, it is assumed that the force would arrive with roughly three days of food, fuel, and ammunition on its vehicles, although if unexpectedly high consumption rates took place aerial resupply might be required prior to three days for one or more of those types of supply.

---

[6] Interview with representatives from the G-4, 82nd Airborne Division, September 11, 2012.

Stryker/LAV Brigade

Due to the large size of the Stryker and LAV-based airborne light armored infantry brigades, the force package is divided into two air-drop echelons and two air-land echelons. The equipment totals listed here assume that the Army would be able to procure various vehicles not in the current inventory. In particular, the Army currently operates no LAV-II–series vehicles. Also, the M8 mentioned here is the light tank that the Army canceled in the 1990s. That vehicle would have to be resurrected for the proposed force structure to be realized, or a different vehicle would need to be substituted, such as the 90-mm assault gun version of the LAV-II or the 105-mm MGS variant of the Stryker. The entire force package is listed in Appendix B. Table 4.1 lists the two airdrop echelons.[7] Note that it includes a division headquarters element as well, in keeping with current practice.[8]

Stryker/LAV Battalion Task Force

The airborne light armored infantry battalion task force is much smaller than the brigade-sized example in Table 4.1, allowing for a single echelon. Table 4.2 lists the vehicles and personnel in this force. Table 4.3 compares vehicle weapons in the LAV-II force versus the Stryker force. In the examples, M-8 light tanks, 105-mm Stryker MGS, or LAV-II systems armed with 90-mm guns would be the vehicles used in assault gun platoons. In addition to those gun-armed vehicles, separate Stryker or LAV TOW-armed antitank vehicles were included in the anti-armor platoons. While the gun-armed vehicles with either 105- or 90-mm guns would be very useful to engage bunkers, buildings, or enemy armor out to a range of roughly 1,500–2,000 meters, for longer-range targets (particularly moving enemy armored vehicles) an antitank guided missile, such as the TOW or Javelin, would be a much more appropriate weapon.

---

[7] Numbers of troops and vehicles were based to the greatest extent possible on current airborne doctrine with Stryker/LAV-type units substituted.

[8] This is based on conversations with XVIII Corps representatives.

Table 4.1
Airborne Light Armored Infantry Brigade
Airdrop Echelons

Element	Number
First Echelon	
Personnel	1,040
LAV/Stryker	99
M8 tanks[a]	9
155-mm howitzers	6
HMMWVs	72
Medium tactical vehicles (MTVs)	37
HEMMTs	4
Trailers (various types)	66
Water trailers	12
UASs	4
Q36 radar	1
Second Echelon	
Personnel	1,137
LAV/Stryker	116
M8 tanks	9
HMMWVs	80
MTVs	34
HEMMTs	19
Trailers (various types)	60
Water trailers	15
M1117s	7
Q37 radar	1

[a] The M8 tank is no longer in production.

Table 4.2
Airborne Light Armored Infantry Battalion Task Force

Echelon	Element
Battalion headquarters	Battalion command section with 1 LAV-C2
	Battalion S3 with 1 LAV-C2
	Signal platoon with 2 LAV-APCs
	Medical platoon with 4 LAV-APCs (ambulance)
	Reconnaissance platoon with 4 LAV-25s
	Mortar platoon with 1 LAV-APC and 4 LAV-M mortar vehicles
Subtotal	106 personnel, including 16 dismounts (reconnaissance scouts), in 17 LAVs of all types
3 rifle companies (each)	Company headquarters with 2 LAV-APCs
	Medical evacuation team with 1 LAV-APC (ambulance)
	Fire support team with 1 LAV-APC (fire support vehicle)
	3 rifle platoons, each with 1 LAV-25 and 3 LAV-APCs
	Mortar section with 2 LAV-Ms
	Assault gun platoons with 3 LAV-AGs (90-mm assault gun)
Subtotal	156 personnel, including 100 dismounts, in 21 LAVs of all types
Brigade-level attachments	
Reconnaissance troop	Troop headquarters with 1 LAV-C2 and 1 LAV-APC
	Medical evacuation team with 1 LAV-APC (ambulance)
	Fire support team, with 1 LAV-APC (fire support vehicle)
	3 reconnaissance platoons, each with 4 LAV-25s
	Mortar section with 2 LAV-M
Subtotal	100 personnel, including 36 dismounts (scouts), in 18 LAVs of all types
Engineer platoon	4 LAV-APCs (engineer variant), 8 crew, and 36 dismounts
Anti-armor platoon	3 LAV-ATs (TOW-II) and 12 crew
Support/ maintenance element	6 LAV-APCs and 18 crew/maintenance soldiers
Battalion task force total	**748 personnel, including 388 dismounts, in 111 LAVs of all types**

SOURCE: This organization was adapted from the current SBCT TOE, as outlined in the following publications: Headquarters, U.S. Department of the Army, *The Stryker Brigade Combat Team Infantry Battalion*, Field Manual 3-21.21, Washington, D.C., April 8, 2003, and Headquarters, U.S. Department of the Army, *Reconnaissance and Cavalry Troop*, Field Manual 3-20.971, Washington, D.C., August 4, 2009.

NOTE: The table was developed to provide a sense of scale for comparison purposes. LAV-APC variants are assumed to be the LAV-IIH model, which, according to General Dynamics Land Systems, has the ability to seat a full nine-person squad. For comparison, an equivalent current Stryker-based battalion task force would have a total of 771 personnel, including 436 dismounts.

Table 4.3
Vehicular Weapons of a LAV-II Force Compared with a Stryker Force

Weapon	LAV-II–Based Force	Stryker-Based Force (current)	Stryker-Based Force (proposed)
Remote weapon stations	54[a]	54	79
Turreted 25-mm Bushmasters	25	25	0
120-mm mortars	12[b]	12	12
Turreted 90-mm cannon	9	0	0
Low-profile 105-mm cannons	0	9	9
Tow missiles	3	3	3

[a] This could be 0.50-caliber, M240, or 40-mm AGL.

[b] Carries fewer rounds than Stryker.

Drop Zone Requirements and Aircraft Operations Profiles

Today the size and number of drop zones required for a given airborne operation depend on the total number of paratroops—and their associated equipment, such as vehicles and heavy weapons—that must be dropped. Compared with today's airborne operations, airborne light armored units might require somewhat larger and/or more drop zones in order to deliver the vehicles so important to this concept.

Low-altitude airdrop would still be used for both personnel and equipment, as is the case today. Air Force transport planes would normally first drop the force's vehicles, then personnel almost immediately after the heavy equipment is delivered, though the decision to drop troops or vehicles first will be made by the ground force commander. The troops would drop either on the same drop zones as the vehicles or on an immediately adjacent location to minimize the amount of time required to link up with their vehicles. Some of the vehicles could be delivered by C-130, while some vehicle types might require C-17s due to their size and weight. Personnel could be dropped by either type of aircraft. Whether C-130s could be used in either case would depend on how far the drop zones are from the airfield(s) where the operation started.

This chapter provided examples of how brigade or battalion-sized airborne light armored infantry units could be organized. These options assume that the force's infantry would be carried in armored personnel carriers from the LAV-II or Stryker families of vehicles, along with other vehicles for direct and indirect fires, plus various support and command-and-control vehicles. Another organizational option that will be explored in separate RAND research is the possibility of having light armor only in battalion-sized units, with the airborne infantry riding in very lightly protected vehicles, such a modified HMMWVs.

Chapter Five examines the airlift implications and requirements of the airborne light armored infantry units that were introduced in this chapter.

Airlift Issues and Requirements

Any new organizational option for U.S. Army airborne forces has to consider the size and composition of the Air Force airlift fleet. This is particularly important for near-term changes to Army airborne forces, since changes to the airlift fleet take place slowly, over time, due to the limited number of new aircraft that can be purchased each year.

Today the Air Force operates three primary transport aircraft, the C-130, C-17, and C-5. The first two aircraft are capable of parachute delivery of troops and equipment, in addition to air-landing them at airports. The C-5 is currently not capable of parachute operations. Additionally, while all the Air Force's C-130 crews are trained and certified to conduct parachute delivery, only a portion of the C-17 force is parachute certified at the time this research was conducted. Army airborne forces regularly train on both types of aircraft.

Delivery Aircraft

The C-17 Globemaster III is the primary heavy airdrop platform for the Air Force. Its mission is "rapid strategic delivery of troops and all types of cargo to main operating bases or directly to forward bases in the deployment area."[1] The C-17 can air-land up to 160,000 lbs of cargo and can airdrop up to 110,000 lbs. It is certified to perform general airdrops from low altitude up to 35,000 feet and airdrop single

---

[1] U.S. Air Force, "C-17 Globemaster III," fact sheet, 2004.

loads up to 60,000 lbs. The normal planning factor for the C-17 is 45 tons of cargo.[2]

This analysis is limited to the C-17 aircraft, though the C-130 Hercules and the C-5 Galaxy offer unique capabilities for this mission. Indeed, the C-130 is by far the most numerous transport plane, with more than 350 in service with the active and reserve components of the Air Force. While the C-130 is not considered an intercontinental, long-range transport plane, Army airborne units typically train on C-130 and size most of their equipment to permit air transport and airdrop from C-130.[3] Appendix C lists the capabilities of these two other Air Force airlift aircraft, including their ability to transport LAV-II and Stryker-class vehicles, as well as personnel. The version of the C-130 that was examined as part of this analysis was the C-130H model, not the newer C-130J.

Limitations and Considerations

Although the specific planning of an airdrop this size is outside the scope of this analysis, it is important to understand certain limitations

---

[2] According to the *Report of the Defense Science Board on Mobility*,

> The maximum dimensional limits of a rigged load (airdrop platform plus energy-dissipating material plus the item to be airdropped plus parachutes) for the C-17 are 118 inches in height, 126 inches in width, and 384 inches in length. The height is further restricted forward of the rigged item's center of gravity to allow extraction under a malfunction condition (that is, if the extraction parachute fails to fully deploy).

> The maximum height for vehicles with rubber tires and vehicles with suspension systems requiring C-17 airdrop is approximately 108 inches. The maximum height for vehicles without suspension systems and for all other equipment is approximately 102.5 inches.

> The C-17's airdrop capability depends on the mode of delivery. The maximum weight that can be airdropped from the C-17 using parachute extraction is 110,000 pounds. The maximum single item that can be airdropped using parachute extraction is 60,000 pounds. As noted earlier, the current maximum rigging capability in the LVAD system is approximately 42,000 pounds. The airdrop hardware presently available can support a single-item maximum gross rigged weight of only 42,000 pounds. This is an airdrop hardware limitation and not an aircraft limitation. The maximum single-item weight for C-17 airdrop, given current 42,000-pound hardware limitations, is about 34,200 pounds, the same as for the C-130. (Defense Science Board Task Force on Mobility, 2005, p. 138)

[3] U.S. Air Force, "C-130 Hercules," fact sheet, December 29, 2011.

of an airdrop this size. These limitations include (1) range; (2) availability of C-17s; (3) the number of airdrop-qualified crews; (4) assembly issues; (5) rigging issues. We discuss these limitations below.

Range

The unrefueled range of the C-17 depends on the load it is carrying. Figure 5.1 depicts this relationship. Essentially, range increases as payload decreases, and, as the figure shows, the increase is linear.

To illustrate the implications of this weight-to-range relationship, here are some examples of potential payloads:

- Airdrop of 102 paratroops (40,000 lbs) = 5,600 nautical miles (nm)
- Airdrop of two LAVs or three MTVs (70,000 lbs) – 4,800 nm
- Maximum airdrop load (110,000 lbs) = 3,600 nm
- Maximum airlift load (170,900 lbs) = 2,400 nm.

Unless the departure airfield is close enough to the drop zone for the heaviest aircraft involved, tanker aircraft will be required to airborne-refuel at least some of the aircraft in the airdrop package. How this issue will be handled will vary considerably, depending on the specifics of an actual operation. For example, transport aircraft might load at Fort Bragg, North Carolina, fly considerable distances to a drop zone(s) with or without refueling, and then land at an airfield much closer to the objective than the original departure air base. Again, the details of a specific operation would decide how this issue was dealt with, including whether there would be a need for in-flight refueling.

Available C-17s

As of the fourth quarter of fiscal year (FY) 2012, there are 180 operational C-17As. The fleet is expected to grow to 204 by the end of FY 2016. The available number of operational C-17s is planned to remain constant at that level from that point on.[4]

---

[4]   We derived the number of available C-17s from the primary aerospace vehicle authorized, which the Air Force defines as the number of aircraft authorized to a unit for the performance of its operational mission.

Figure 5.1
C-17 Range Versus Load

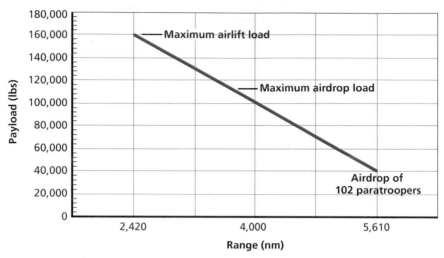

SOURCE: Data from Boeing Corporation, 2014.
RAND *RR309-5.1*

Airdrop-Qualified Aircrews

Currently only about 10 percent of C-17 crews—fewer than 100 total—are qualified for airdrop operations. On any given day, these crews are deployed worldwide on operational missions. In addition, crew rest requirements could further limit the availability of these crews, especially for a mission that involves multiple echelons of the airborne force, thus requiring several sorties by each airdrop-qualified aircrew in order to complete the delivery of the Army airborne units.

Assembly Issues

Extensive planning would be required to bring together the aircraft with the airdrop equipment and personnel, get the equipment rigged and loaded, then position the aircraft at a staging base to await the mission. Ramp space at airfields limits the maximum number of aircraft on the ground at any one time (referred to as "maximum operating on the ground"). With the numbers of C-17s in these airdrop options, it is highly likely that no single staging base will suffice, requiring sequencing the loading of aircraft, then staging them at different airfields prior to the mission.

Rigging Issues

Rigging this number of vehicles for airdrop would require considerable time and effort, especially for brigade-size operations. Indeed, as of late 2012, the Army's airborne units probably lacked the capacity to rig this number of vehicles in time to meet rapid-response goals, such as the requirement for a battalion task force from the 82nd Airborne Division to depart from Fort Bragg within 24 hours of alert. Linking the riggers with the equipment, then co-locating the rigged equipment with the aircraft, will add additional complexity. Rigging facilities are also a limiting factor. Consideration should be given to rigging at several bases in parallel to cut down the required time to rig this large and complex number of vehicles and equipment. This too, would present significant logistical challenges. It may be possible to maintain a portion of the unit equipment in a prerigged status to reduce rigging time. Prerigging of equipment is a technique used today by Army airborne units. It should also be noted that today the 82nd Airborne Division does not have an organic rigging capability—that is provided by XVIII Corps assets.

Analysis: Stryker/LAV-Based Airborne Light Armored Infantry Brigade

The following tables describe the airlift requirements for echelons 1 and 2 for an airborne light armored infantry brigade. Is it assumed that planners would want to airdrop echelons that are heavily biased toward combat elements of the BCT. It is assumed that the remaining two echelons of the BCT would arrive later, in an air-landing mode, and would include the majority of the brigade's logistics elements. There is a base case and an excursion case. The base case includes some changes to the Stryker TOE because some of the vehicles, namely the HEMTT, are oversized and would require modifications to adapt them for airborne operation. The dimensions, weights, and numbers of vehicles that can be airdropped by C-17 are listed in Appendix D. The excursion case includes steps taken to decrease the number of required C-17s for the operation, including the following:

- Dual load troops with equipment (will require two passes over the drop zone, equipment dropped first, then personnel).
- Substitute M119 (105 mm) for M777 (155 mm).
- Substitute 50 percent of MTVs/trailers with M1114s/1.5-ton trailers.
- Delete bridging equipment.
- Delete M8s (or Stryker MGS or LAV-II 90 mm) and rely on airpower for dealing with any enemy armor encountered.

The first column of Tables 5.1 and 5.2 show the personnel and equipment included in each echelon. The second and third columns show the number of C-17s required to lift the personnel or equipment. The excursion case applies the steps listed above. The fourth column shows the difference in number of C-17s required between the base and excursion cases. Summing across the echelons for the Stryker-based version shows that the excursion requires 57 fewer C-17s than the base case. The savings in aircraft result from loading troops on the same aircraft that carry equipment, which requires ten fewer C-17s. Additional savings result from substituting equipment, substituting 1.5-ton trailers for MTV trailers, and replacing 155-mm howitzers with 105-mm variants. Other savings come from deleting equipment: bridging equipment and M8 tanks.

Table 5.1 shows the results of substituting LAVs for Strykers. As was the case in the Stryker example, the excursion case for the LAVs requires fewer (57) C-17s than the base case. However, it also takes 36 fewer C-17s to move the LAVs. The Stryker brigade required 245 C-17 sorties, and the LAV brigade only 209.

Analysis: Stryker/LAV-Based Airborne Light Armored Infantry Battalion Task Force

Tables 5.3–5.5 show C-17 aircraft requirements for the airborne light armored infantry battalion task force. Again, we present a base case and an excursion case. The base case includes some changes to the

Table 5.1
Stryker-Based Airborne Light Armored Infantry Brigade Airdrop Echelons

Items	Number	Echelon 1			Number	Echelon 2		
		Base Case	Excursion Case	Base Case Compared with Excursion		Base Case	Excursion Case	Base Case Compared with Excursion
Personnel	1,040	10	0	-10	1,137	11	0	-11
Strykers	99	50	50		116	58	58	
HMMWVs	72	9	9		80	10	10	
HMMWV trailers	50	6	6		43	6	6	
MTVs	37	18	12	-6	34	17	11	-6
MTV trailers	12	6	4	-2	17	8	5	-3
Water trailers	9	2	2		15	3	3	
M-8 tanks	9	5	0	-5	9	5	0	-5
Howitzers	6	3	1	-2	7	2[a]	2	
UASs	4	0	0		N/A			
Bridging	2	2	0	-2	N/A			
MTVs for HEMTTs	4	2	2		19	10	6	-4
Q36/Q37 radar	1	0	0		1	2[b]	1	-1
Total C-17s required		113	86	-27		132	102	-30

[a] 105-mm howitzer. [b] Q37 radar.

Table 5.2
LAV-Based Airborne Light Armored Infantry Brigade Airdrop Echelons

Items	Echelon 1				Echelon 2			
	Number	Base Case	Excursion Case	Base Case Compared with Excursion	Number	Base Case	Excursion Case	Base Case Compared with Excursion
Personnel	1,040	10	0	−10	1,137	11	0	−11
LAVs	99	33	33		116	39	39	
HMMWVs	72	9	9		80	10	10	
HMMWV trailers	50	6	6		43	6	6	
MTVs	37	18	12	−6	34	17	11	−6
MTV trailers	12	6	4	−2	17	8	5	−3
Water trailers	9	2	2		15	3	3	
M-8 tanks	9	5	0	−5	9	5	0	−5
Howitzers	6	3	1	−2	7	2[a]	2	
UASs	4	0	0		N/A			
Bridging	2	2	0	−2	N/A			
MTVs for HEMTTs	4	2	2		19	10	6	−4
Q36/Q37 radar	1	0	0		1	2[b]	1	−1
Total C-17s required		96	69	−27		113	83	−30

[a] 105-mm howitzer. [b] Q37 radar.

Table 5.3
C-17 Sorties Required for Stryker and LAV-Based Airborne Light Armored Infantry Battalion Task Forces

Items	Number	Echelon 1			Echelon 2		
		Base Case	Excursion Case	Base Case Compared with Excursion	Base Case	Excursion Case	Base Case Compared with Excursion
Crew	302	3	0	−3	3	0	−3
Dismounts	384	4	0	−4	4	0	−4
Command vehicles	3	2	2		1	1	
Personnel carriers	37	18	18		12	12	
Ambulances	8	4	4		3	3	
Reconnaissance/25-mm vehicles	25	12	12		8	8	
Mortars	12	6	6		4	4	
Fire support	4	2	2		1	1	
Assault guns	9	5	5		3	3	
Engineer vehicles	4	2	2		2	2	
Antitank vehicles	3	2	2		1	1	
Total C-17s required		60	53	−7	42	35	−7

Table 5.4
C-17 Sortie Requirement for Stryker and LAV Brigades

Force Option	Base Case	Excursion Case	Base Compared with Excursion
Stryker brigade	245	188	−57
LAV brigade	209	152	−57
Stryker battalion	60	53	−7
LAV battalion	42	35	−7

Stryker TOE to make it a solely Stryker- or LAV-based force. The only change for the excursion case is the dual loading of troops with equipment, which would require two passes over the drop zone: one to drop the equipment and a second to drop the troops.[5] Dual loading reduces the requirement for aircraft.

Summary

The brigade-sized force options, for both the Stryker and the LAV options, require almost the entire operational C-17 fleet to successfully accomplish the airdrop. The battalion-sized task force, though requiring a significant portion of the operational C-17 fleet, can be achieved with existing aircraft and trained airdrop crew resources. The LAV-based force is the most realistic option considering the myriad limitations on resources and logistics.

This new concept would provide airborne forces with increased mobility, lethality, and survivability. Tactically, a fully motorized force could conduct mounted maneuver to allow greater offset from the point of entry to the objective. This would provide several advantages. For example, (1) airborne forces could avoid some of the adversary's air defense threats, (2) the adversary may not be able to identify the specific objective quickly, and (3) therefore, it could force the adversary to defend a larger area.

---

[5]   Note that these drop packages do not account for a BCT headquarters element.

Most importantly, this new concept could provide decisionmakers with new strategic options to stabilize potential conflicts more quickly and prevent them from escalating and could prevent conflicts from escalating. It could also provide increased flexibility in utilizing the Global Response Force. As a result, this new concept would enable airborne forces to be rapidly deployed and play a key role in a broader set of missions. Next, we discuss these potential mission sets.

CHAPTER SIX

Potential Uses for Airborne Light Armored Infantry Forces

The proposed airborne light armored infantry force enjoys several advantages over traditional airborne infantry in three domains: It has substantially improved tactical mobility, it has improved lethality against a range of targets, and it is more survivable against a number of threats. These advantages—which come at the cost of additional airlift—could make the force well suited for employment in a range of operational contexts.

In this chapter, we discuss how an airborne light armored infantry formation might be employed in seven different vignettes and one new role. These represent possible circumstances in which the force might be used and are drawn from historical experience and plausible future contingencies. For each case, we describe how the notional unit's improved tactical mobility, firepower, and protection would enhance its capabilities versus those of existing airborne infantry units.

In each of these vignettes it was assumed that sufficient transport aircraft were available, and within range of the drop zones, to deliver an airborne light armored infantry brigade combat team or battalion task force appropriate for the mission. This might require several sorties for the transport aircraft, particularly for a brigade-sized force. There would also have to be sufficient rigging capability either at the point of departure of the airborne force (e.g., Fort Bragg, North Carolina, or northern Italy) or at an intermediate staging base.

In the event of an actual crisis similar to a situation described in the scenarios in this chapter, a variety of considerations would influence planning, including the amount of time available to respond, the severity of the threat, and the amount of airlift available to deploy the Army airborne force. In some cases the planner might desire that an entire brigade be deployed, while in other situations a battalion-sized element would be adequate, at least initially. The challenge with a brigade-sized force would be the large number of aircraft that would be required to deploy and sustain the now more highly motorized airborne force. As a minimum, a brigade-sized unit would require multiple sorties by the available transport planes to deploy the unit.

Vignette 1: Counter Genocide

The 1994 Rwandan genocide involved atrocities committed throughout the country, in a remote and underdeveloped region of sub-Saharan Africa far from any coastline. Over a period of nearly two months, extremists from the nation's Hutu ethnic majority murdered between 500,000 and 800,000 of their countrymen, mostly ethnic Tutsis. The small United Nations Assistance Mission for Rwanda (UNAMIR) was hamstrung by its rules of engagement and outgunned by the Hutu-dominated Rwandan military, or Rwandan government forces. UNAMIR's helplessness was grotesquely exposed when Rwandan soldiers murdered the country's prime minister, torturing and killing the ten Belgian peacekeepers who had been sent to protect him.[1]

Any intervention force would have needed to respond rapidly, deploy quickly, have sufficient tactical mobility to deal with developing events across an area a little smaller than Massachusetts, and overmatch any resistance from either the Rwandan government forces or the Rwandan Patriotic Front, a rebel force which exploited the chaos of the genocide to invade and overrun the country. Widespread combat between the two groups erupted shortly after the genocide began and constituted the context within which any intervention would have been conducted.

---

[1] The best description of the Rwandan genocide is probably Gourevitch (1998). Dallaire (2003) tells the frustrating and often horrifying story of UNAMIR from the perspective of its commander.

U.S. Army Colonel Scott Feil wrote an influential essay describing "how the early use of force could have succeeded in Rwanda." He argued that a brigade-sized infantry force could have moved quickly into the country and "shut down . . . acts of violence."[2] An analysis of the deployment requirements for such a force, however, suggests that even under optimistic assumptions about the ability of regional airports to accommodate airlift operations, it would have taken in excess of two weeks—and up to a month or more under more realistic conditions—to close a force of this size and configuration into Rwanda.[3] Much of the slaughter would, in other words, have already occurred by the time the intervention force was in place and prepared for operations.

A substantial amount of the airlift required to move the force would have been consumed by the helicopter assets Feil assumed would be needed to provide tactical mobility to the force. The intrinsic mobility of an airborne light armored infantry force would have reduced the requirement for rotary-wing transports and allowed the rapid insertion of a force able to defeat any resistance it was likely to encounter from an indigenous force anywhere in the country. Follow-on airdrops could have been used to jump traditional airborne infantry units to secure areas cleared by the more heavily armed, armored, and mobile motorized force.

The deterrent effect that even a small, capable force might have exerted should probably not be underestimated. One participant in a conference on Rwanda argued, "It was only when the extremist perpetrators sensed that the world was not going to address the crisis and that UNAMIR's contingents were in a self-protection mode that the genocide began in earnest."[4] The mere knowledge that a small but

---

[2]   Scott R. Feil, *Preventing Genocide: How the Early Use of Force Might Have Succeeded in Rwanda*, New York: Carnegie Corporation, 1998, p. 13.

[3]   The analysis, based on U.S. Army and Air Force planning factors, concluded that approximately 297 C-141 and 60 C-17 sorties would have been required to deploy a custom-configured task force built around a reinforced air assault infantry brigade. See David A. Shlapak, John Stillion, Olga Oliker, and Tanya Charlick-Paley, *A Global Access Strategy for the U.S. Air Force*, Santa Monica, Calif.: RAND, MR-1216-AF, 2002, Chapter 4, especially pp. 73–84.

[4]   Feil, 1998, p. 39.

capable force was on its way to Rwanda might have spared the lives of hundreds of thousands of innocents.

Vignette 2: Establish a Deterrent Presence

One potential use for an airborne light armored infantry force would be to rapidly establish a deterrent presence on the ground in the midst of an urgent crisis. This is not an unfamiliar role for paratroopers. As mentioned in Chapter Two, in 1990, a brigade of the 82nd Airborne was deployed to Saudi Arabia in response to Iraq's invasion of Kuwait. Air-landed at Dhahran Airfield over a period of about five days, the brigade's mission was—according to the official Army history of the Gulf War—to establish a "toehold" on Dhahran and the port at al-Dammam.[5] "The tactical situation," says that history, "was tenuous in the extreme."[6] In combination with air power that was also deploying into the region, the hope was that the force would persuade Saddam Hussein not to press his attack into Saudi Arabia.

While the airborne brigade sent to Saudi Arabia was large—almost 4,600 soldiers—it possessed little tactical mobility and a minimal amount of armored capability: a single company of M551 Sheridan light tanks. Its tank-killing capability, which would have proved vital in any fight with Iraq's heavily armored forces, resided primarily in the battalion of attack helicopters that deployed with it. Movement of that battalion accounted for many of the 250 C-141 sorties and the five and a half days it took to move the brigade. Furthermore, although the brigade did include that small number of Sheridans, the infantry was still limited to walking, since they had no armored personnel carriers or even trucks. Later, the early-deploying airborne units purchased commercial trucks from local Saudi dealers to give them some degree of mobility, which was, of course, totally unarmored.

---

[5]   Robert H. Scales, Jr., *Certain Victory: The U.S. Army in the Gulf War*, Washington, D.C.: Potomac Books, 2006, p. 85.

[6]   Scales, 2006, p. 85.

An airborne light armored infantry force of the kind described in this report would be valuable as a leading-edge element of a similar deterrent response or as the main body of one in a smaller-scale context. In a deterrent response role, an airborne light armored infantry battalion could deploy more rapidly by being airdropped, thereby also avoiding reliance on a fixed air base that could be targeted by an adversary armed with the kind of modern long-range precision strike weapons that Iraq lacked in 1990. Were air-landing possible, as it was in 1990, an airborne light armored infantry force would provide a degree of firepower and protected mobility that the paratroop infantry of the 82nd lacked in Desert Shield.

While any small force would be very dependent on external support—principally air power—for tactical viability against an opponent fielding a substantial amount of heavy armor, the airborne light armored infantry force would have the advantages of mobility, improved survivability against indirect fires, and some amount of protected anti-armor firepower over a classical airborne infantry force. Such a formation could put combat power close to that of a traditional airborne brigade on the ground faster and do so while putting many fewer paratroopers at risk.[7] Had the enemy attacked in overwhelming numbers, a LAV-armed airborne force, if available, could have conducted a mobile delaying action, inflicting casualties on the advancing Iraqis as it withdrew southward. In comparison, a foot-mobile airborne infantry unit would have been at great risk of being pinned in place, encircled, and annihilated.

Vignette 3: Protect an Enclave

In April 1993, the United Nations declared the Bosnian town of Srebrenica a "safe area" and put it under the protection of an improvised battalion of Dutch infantry. In July 1995, troops from the

---

[7] We did not undertake the kinds of combat simulation that would enable us to make definitive comparisons between the capabilities of various force configurations in this or any other scenario. Such analysis would be a very useful and informative extension of these initial assessments.

Serbian Army of Republika Srpska (VRS) overran the town with minimal resistance from the peacekeepers, who were outgunned by the Serbs' artillery and small amount of heavy armor and proved unable to rely on effective NATO air support. In the aftermath of Srebrenica's fall, Serb forces massacred more than 8,000 Bosnian Muslim men and boys, an event that then–UN Secretary-General Kofi Annan characterized as the worst crime perpetrated in Europe since World War II.[8]

The location and time scale of the events in Srebrenica afforded ample opportunity for the international community to deploy a conventional force capable of defeating the VRS forces that besieged and attacked the town—had there been the will to do so. Under other circumstances, however, a more rapidly moving ethnic conflict or one taking place farther from existing U.S. bases could call for an intervention force simultaneously capable of being quickly dispatched to a distant, perhaps remote, location and, with joint support, defeating an adversary equipped with some tanks and other heavy weapons. An airborne light armored infantry force would be suitable on both scores.

The attack on Srebrenica unfolded over a period of five days, July 6–11, 1995. It is possible that, depending on the alert status of the unit, a light armored infantry battalion task force could have been inserted into the enclave entirely during that period. As in the Rwandan case, the deterrent effect of knowing that such a force was en route might have altered Serbian calculations regarding the advisability of their actions.

In 1995, Srebrenica's defenders found themselves in the debilitating tactical situation that their adversary, the VRS, controlled all land lines of access to the enclave. The Serbs prevented resupply and reinforcement of the Dutch peacekeepers, even forbidding soldiers who had gone on leave from returning to their units. An airborne unit could have bypassed VRS roadblocks and been resupplied by air, nullifying this significant Serb advantage.

---

[8]   The Srebenica-Potocari Memorial Center (undated) lists 8,373 victims of the massacres in the Srebrenica enclave. As of July 2012, the remains of more than 7,000 victims had been accounted for (ICMP, 2012). For Annan's remarks, see United Nations, 2003.

Traditional airborne infantry units could also have deployed rapidly to the area and operated independently of ground lines of communication. They would not, however, have possessed the tactical mobility, anti-armor firepower, and protection against artillery fire that a LAV-based force would enjoy. All of these characteristics would have proven immensely valuable in the defense of the men, women, and children gathered into the Srebrenica enclave.

Vignette 4: Seize and Secure a WMD Site

The WMD elimination (WMD-E) mission is one of the most challenging that the military could confront. U.S. forces could be asked to secure one or more key sites as part of an ongoing conflict with an adversary, inside a collapsing or failed state, or as an independent operation aimed at eliminating a specific threat.

For most likely adversaries, WMD sites—especially any that might be related to nuclear weapons—would probably be heavily protected by both air and ground defenses. The proposed airborne light armored infantry force's ability to be inserted at some distance from the target, hence avoiding the defenses concentrated in its immediate vicinity, would be an advantage in such circumstances. Once on the ground, the airborne light armored infantry troops would be able to maneuver rapidly (depending on terrain or weather constraints) to survive contact with and defeat a wider range of opposition than could today's generally foot-mobile airborne forces.

To illustrate this, consider an operation to secure the North Korean nuclear installation at Yongbyon. In the event of a North Korean collapse or regime implosion, the status of the country's nuclear materials and facilities would likely be of enormous concern to the United States and its allies, and substantial risks might be taken to ensure that they were secure. As Figure 6.1 shows, the facility is close to both the North Korean coastline and two air bases, perhaps permitting a multiazimuth joint forced-entry operation. In other contexts, airdrop may be the only viable option; most of Iran's nuclear facilities, for example, are located much too far inland to be reached by amphibious forces.

Figure 6.1
Yongbyon Concept

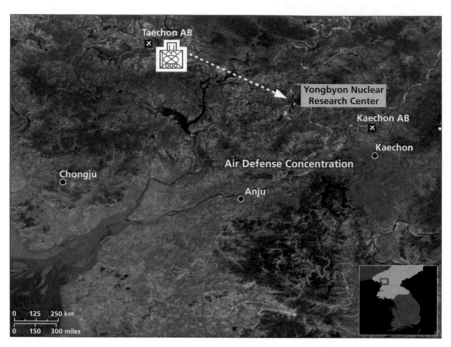

RAND *RR309-6.1*

Note the dotted red ring around Yongbyon; this indicates an area containing numerous air defense sites—for both missiles and AAA— clearly intended for defense. Because of their inherently limited mobility, existing airborne infantry formations would need to jump into or very near to this heavily defended area to avoid a prolonged and potentially debilitating march to the objective. This would increase the risk to the lift aircraft, which would have to fly low and slow over or near the most likely places for concentrations of difficult-to-suppress MANPADS and optically guided AAA. Furthermore, compared with an airborne light armored infantry force, a foot-mobile infantry unit would be at a higher risk of being pinned down by enemy mortar or artillery, or even intense small-arms fire, thus slowing or stopping its advance toward the time-sensitive objective.

Infantry would also be at a disadvantage against the kinds of forces likely defending as valuable a facility as Yongbyon, which could include both heavy and light armor. An airborne light armored infantry force would offer more protection for the troops, increased anti-armor firepower (especially the mobile gun system and TOW-armed variants), and the ability to maneuver to avoid threats or put opposing forces at the tactical disadvantage.

It should be noted that for a light armored airborne infantry force to be viable in a scenario like this that includes an opponent with a large number of armored vehicles, including main battle tanks, certain conditions would have to be met. If the North Korean armed forces were in the process of disintegrating due to, for example, an internal civil war and their units were no longer considered cohesive, that might provide sufficient opportunity to deploy this type of U.S. airborne force, since the likelihood of encountering cohesive enemy armored forces would likely be low. Alternatively, if the North Koreans were still operating as cohesive units, it might be possible to apply joint fires (mostly from the Air Force and Navy) to seal off the immediate areas of operations from North Korean armored units. In any case, the possibility of encountering enemy armor would be a very major consideration in this type of situation.

Vignette 5: Conduct a Noncombatant Evacuation Operation

NEOs are unpredictable in timing, size, and nature. While most can be conducted with the cooperation, or, at minimum, the acquiescence, of the host government, the U.S. military must be prepared to execute them in circumstances where some degree of resistance—even organized resistance by elements of a state's military—may be encountered. A NEO of this type would resemble a modest-sized forced-entry operation and could be an appropriate contingency for an airborne light armored infantry capability. The force concept depicted in this report would provide a degree of protection and mobility that could be critical to safely extracting U.S. citizens from harm's way.

Nigeria's energy infrastructure is principally located in the Niger River Delta, with the town of Port Harcourt serving as its principal oil terminal. Nigeria's capital, Abuja, is 470 km from Port Harcourt, while the country's most populous city, Lagos, is 540 km from Abuja and 440 km from Port Harcourt. Together, the three cities are the vertices of a triangle covering almost 100,000 sq km, shown in Figure 6.2.[9] With political and ethnic unrest endemic in the Niger Delta and beyond, the possibility that U.S. and other third-country nationals might require evacuation from all corners of this area is far from improbable.

In such a scenario, both the strategic and tactical mobility of an airborne light armored infantry force would prove valuable. The

Figure 6.2
Nigeria

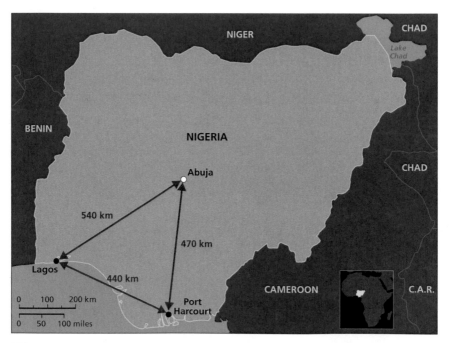

RAND RR309-6.2

---

[9]   For purposes of comparison, Indiana covers about 94,000 sq km, while Kentucky is about 105,000 sq km in size.

former would allow it to be inserted rapidly wherever it would be most advantageous; entry would not necessarily be tied to an air base or seaport. The latter would permit it to speedily traverse the distances that might be necessary to gather up dispersed foreign nationals and transport them to a secure departure location. The firepower and protection offered by the unit's light armored vehicles would help make the entire operation viable in a situation where some degree of resistance was expected or feared.

Vignette 6: Conduct a Humanitarian Assistance and Disaster Relief Operation

HADR operations are among the most common taskings for U.S. military forces. Although many are fairly small, straightforward, or both, more challenging ones are plausible. One such scenario would involve a severe earthquake in the Turkey-Iraq-Iran border region. Damage to the area's transportation infrastructure could make it difficult to move relief personnel and supplies by land, violence between Kurdish separatists and government forces—or among different Kurdish factions—could make for a moderately hostile operational environment, and distance from salt water would make entry from the sea challenging. At the same time, helping provide prompt relief to those affected by the disaster could be important in maintaining the credibility of friendly governments and sustaining political stability in the region.

This is another scenario in which the size of the potential area of responsibility would make motorization a distinct advantage over a foot-mobile force. As in the NEO case, a battalion-sized light armor task unit could provide the protected, mobile firepower that enables a larger, softer unit to operate in relative safety. An adversary fielding no weapons heavier than mortars or RPGs would be overmatched by the airborne light armored infantry force, which would also offer improved effectiveness (versus heavier resistance) over that of a traditional airborne unit. By screening convoys, eliminating enemy strong points, or providing quick-reaction forces to help protect relief centers and refugee camps, strategically deployable airborne light armored infantry capabilities would permit both military and civilian entities to begin providing assistance quickly in circumstances in which they might otherwise be at too great a risk to operate effectively, if at all.

Vignette 7: Airborne Light Armored Infantry Forces in State-to-State Conflict

Airborne forces have seen heavy use in recent U.S. wars, but not in a primarily airdropped role. Instead, the 82nd Airborne Division was reinforced and employed as an infantry unit in the major combat phase of the invasion of Iraq, and both it and the Army's other airborne brigades have been tapped to provide troops for rotational missions in both Iraq and Afghanistan.

However, as mentioned in Chapter Two, on March 23, 2003, the 173rd Airborne Brigade conducted one of the largest airborne assaults since WWII when 954 paratroopers jumped onto Bashur airfield in northern Iraq. Over the subsequent three-plus days, the remainder of the brigade was air-landed at the base, and the 173rd subsequently conducted operations aimed at eliminating Iraqi forces and securing the key oil center of Kirkuk, which fell in early April. While the airdrop was considered a combat assault, little resistance was anticipated, and none was encountered.

Even after the bulk of the 173rd Airborne Brigade arrived in northern Iraq, there was hesitation about advancing southward. Although at that time the 173rd was the most heavily motorized U.S. Army airborne unit (the two-battalion 173rd had a considerable number of HMMWVs), there was no willingness to advance until armor arrived from U.S. units in Germany. It took considerable time to deploy a platoon of Abrams main battle tanks and another platoon of Bradley infantry fighting vehicles to northern Iraq by air.

Had the airborne force in northern Iraq been armed with several dozen LAVs or Stryker vehicles, it is possible that there would have been an ability—and willingness—to advance sooner.[10] The fact that LAVs or Strykers would be easier to deploy by air (including using a parachute option) would have meant that combat power would have been built up more quickly than was the case in 2003, when considerable time was required to fly in platoon-sized elements of Bradley and Abrams vehicles. This is an example of how a future airborne unit with greater mobility, protection, and firepower could have had a more significant effect on a conventional combat operation than the generally foot-mobile airborne forces of 2003—or today.

If opposition were expected to be heavier, an airborne light armored infantry force might have been a better option than light infantry. In addition to the armored force being better able to survive and defeat a wider range of threats, the mobility offered by an airborne light armored infantry force would have enabled more rapid movement to, for example, establish a perimeter around the base. This mobility, combined with the unit's superior organic firepower, would have offset the relatively smaller number of rifles available to the airborne light armored infantry force versus a traditional airborne infantry battalion. A few additional planeloads of paratroopers, either airdropped or quickly air-landed, could fully make up that deficit, leaving the airborne light armored infantry force free to expand the lodgment's footprint, conduct reconnaissance of the surrounding area, or take advantage of the element of surprise to begin attacks on enemy forces.

New Role: Airborne Cavalry

An airborne light armored infantry force could also represent a whole new class of capability for the U.S. military—that of very rapidly deployable cavalry. In some ways, this role would parallel one of the primary ones envisioned by Soviet planners for their mechanized airborne forces: "to support the rapid advance of a large combined arms force

---

[10] RAND interviews with the staff of the 173rd Airborne Brigade, Vicenza, Italy, summer 2005.

deep into the enemy's operational or operational-strategic depth."[11] Traditional airborne forces' lack of mobility once on the ground clearly renders them unsuitable for such a mission. Marine reconnaissance battalions, while capable, are no easier to project hundreds of kilometers inland than are medium-weight Army forces of comparable size. Only an airborne light armored infantry force could offer the combination of strategic and tactical mobility needed to carry out such a mission.

Employed in this role, an airborne light armored infantry unit could provide intelligence about enemy strengths and dispositions, engage in economy of force offensive operations, disrupt the opponent's rear area, screen the movement of other U.S. forces, divert defenders from the main effort, and compel the enemy to move in response to its presence, in the process exposing enemy forces to devastating attacks from U.S. air power.

Concluding Thoughts

None of these vignettes or roles can be said to decisively establish the superiority of the proposed airborne light armored infantry force over all alternatives, including traditional foot-mobile airborne infantry. Each could have been crafted in a way that made it appear either more or less amenable to the employment of various kinds of forces. However, four overall points seems worth making.

First, it is not the proposed force's strategic mobility, tactical mobility, firepower, or level of protection taken individually that offers advantages over alternative formations, but combinations of the four. These characteristics carry different weights in different circumstances, but no existing or proposed unit configuration appears to offer the flexibility conferred by an airborne light armored infantry force across the range of "medium-intensity" scenarios presented. A Stryker battalion is obviously the closest analog in the Army order of battle, but it requires a secure air base on the debarkation side of any deployment. Marine

---

[11] Headquarters, U.S. Department of the Army, *The Soviet Army: Specialized Warfare and Rear Area Support*, Field Manual 100-2-2, Washington, D.C., July 16, 1984, p. 2-1.

amphibious battalions offer many capabilities similar to those of the proposed force, but their utility diminishes fairly rapidly as the locus of operations moves away from the littoral or if the Marine Air-Ground Task Force is far from the crisis area.

Second, in addition to improved operational effectiveness, the attributes of an airborne light armored infantry force could make it a more strategically attractive crisis option for decisionmakers. All three of its core qualities combine to reduce risk to the soldiers in the unit: Its firepower helps it defeat a wider range of threats, its protection physically shields troops from harm, and its mobility allows the force to avoid danger or maneuver to engage an adversary from a more advantageous tactical position. Its advantages over existing forces also make it more adaptable to unanticipated circumstances, such as encountering an unexpected level of resistance. While no military operation is safe, and certainly not an airborne or other forced-entry operation, political leaders may find it easier to employ, or gain political leverage by credibly threatening to employ, a force that is substantially less vulnerable to likely threats than a traditional airborne infantry. This may be especially the case in situations like atrocity-response scenarios, in which concerns over friendly losses can strongly influence the willingness of decisionmakers to commit troops.

Third, the value of an airborne light armored infantry capability should be assessed not just in isolation but also according to its potential role as a leading-edge or escort force for other units responding to a contingency. For example, a battalion-size task force could provide security for a much larger HADR operation or NEO. It could also serve as a kind of rapidly deployable cavalry or diversionary force for a larger-scale conventional attack in a more mainstream combat contingency. Traditional airborne infantry would not be particularly appropriate in either role.

Finally, any light- or medium-weight force injected into a hostile environment will be highly dependent on joint capabilities for its effectiveness, if not its survival. It will remain impossible to rapidly move the kinds and sizes of ground combat forces that guarantee overmatch, regardless of the tactical situation. Instead, strategically mobile units will continue to rely on firepower support and ISR supplied primarily by air and space systems.[12] Employing an airborne light armored infantry force will not eliminate this dependency; the volume of aerial resupply required compared to a traditional airborne infantry unit will greatly increase. But the advantages such a force offers in firepower, mobility, and protection should make it more usable with somewhat less support than would be the case with existing force configurations. The severity of the threat, in terms of air defenses, long-range surface-to-surface fires, and the opponent's ground maneuver forces, would all influence whether the type of enhanced airborne force described in this report would or could be employed in a particular situation.

---

[12] Some of the air capabilities will of course be projected from maritime platforms, and under some circumstances, sea-based capabilities, such as naval gun support, may directly be brought to bear.

CHAPTER SEVEN

Issues Related to the Implementation of the New Concept

As the vignettes in Chapter Six illustrate, the potential advantages of this new concept include enhanced mobility, lethality, and survivability. However, these enhancements would not come without costs and trade-offs. In this chapter, we lay out some of the issues related to the implementation of this concept, including potential barriers the Army would need to overcome to implement it, the implications this concept would have on the joint force, and organizational options for implementation.

Issues the Army Would Need to Resolve

Funding issues will likely be one of the main barriers to implementing this new concept, especially as defense budgets shrink. However, LAV-II and the Stryker are already in the U.S. military inventory; therefore, costs would be lower than if a new vehicle were to be developed and purchased. As mentioned earlier in this report, the advantage of the Stryker is that it already exists in the Army inventory (although some additional versions would be needed) and is a proven vehicle that the Army is comfortable with. The advantage of the LAV-II is that the vehicle is better suited to the airborne (including airdrop) mission, particularly in terms of the C-130 aircraft.

In addition to equipment costs, the Army would also need to consider the costs of training with the new equipment and new tactics, techniques, and procedures (TTPs). For instance, some Army airborne personnel would need to be trained on the concept's various vehicles and associated TTPs. As mentioned earlier, this concept could entail relying on JPADS-delivered resupply for motorized elements advancing toward the objective. These changes in airborne operations would be quite a departure from traditional airborne TTPs and would require a number of changes on the part of the airborne community. Additionally, the Army would need to expand its rigging capacity, particularly its ability to rig vehicles much larger and heavier than the HMMWVs currently used by airborne forces. Finally, the Army would need to determine the costs associated with sustaining an airborne light armored infantry force.

Implications of This Concept for the Joint Force

The implementation of this concept would also have implications for the joint force. While we cannot provide an exhaustive list, as indicated by our analysis in Chapter Five, this concept would generate additional airlift requirements for the Air Force (though it would not require the Air Force to procure additional aircraft), since an Army airborne unit with larger numbers of vehicles would need more aircraft for transport. Additional training would also be needed for Air Force aircrews. The greater number of transport aircraft required by this concept could require that a greater percentage of C-17 aircrews be airdrop-capable than is the case today. If the C-5's airdrop capability could be increased, it could greatly add to existing airdrop capacity without requiring new airframes. To do so would involve significant training requirements, however, since no C-5 crews are currently trained for airdrops. The costs associated with sustaining a modified airlift fleet will also need to be identified.

In addition, the joint force will be needed to support this airborne force and enhance its mobility. For example, joint fires and close air support will be needed to protect the airborne force. Offensive counter-air, suppression of enemy air defenses, airborne command and control, and tanker support would normally be required to support an airborne operation. In addition, joint ISR capabilities will be needed to facilitate the force's mobility.

In terms of the logistics requirements of this type of light armored airborne infantry force, it has already been mentioned that JPADS might be a particularly attractive way of delivering resupply to a maneuvering airborne unit that now includes a far larger number of vehicles than is the case today. To operationalize this resupply concept, joint training would be required for Army light armored airborne infantry units and the Air Force, with various types of transport planes delivering JPADS pallets in various terrain types and weather conditions. According to an Air Force representative familiar with airdrop crew qualifications, a C-17 aircrew that is required to drop JPADS must go through extensive training that differs from that required to airdrop personnel. For example, Air Force loadmasters treat JPADS packages similarly to the containerized drop system packages, and both procedures require similar training for both pilots of an aircraft; however, for a JPADS mission, one pilot requires extensive training to operate the JPADS computer system. That requirement could impose an additional burden on the Air Force's training programs. But if it were met, there would be an increased number of aircrew members available for both JPADS and personnel/heavy equipment airdrop.

Organizational Options for Implementation

The Army has several near-, mid-, and long-term options for implementing the new airborne concept. In the short term (over the next five years), one option it might consider is to experiment with existing vehicle options for this concept. For instance, the Army could enter into an agreement with the Marine Corps to test whether LAV-25s might be a feasible vehicle for this concept. This experimentation would enable the Army to determine vehicle requirements and numbers and, working with the Air Force, explore the airdrop ramifications. Another option would be to combine combat vehicles (Stryker or LAV-II) with HMMWVs or similar vehicles. This near-term option has advantages in that no new equipment would need to be developed, and the existing airlift fleet would be used. In addition, this experimentation would allow the Army to more clearly determine its own requirements, as well as changes in joint capabilities that would be needed to support the full implementation of the new airborne concept.

In the mid-term (four to ten years), the Army could identify and develop any necessary new airdrop capabilities. For instance, it could take steps toward developing a heavy drop capability above 42,000 lbs or a JPADS capability up to 42,000 lbs.

In the long term, the Army could fully implement the concept by developing and fielding vehicles specifically designed with superior characteristics for this mission set, particularly airdrop. This would entail the widespread use of combat vehicle airdrop, improved forms of personnel delivery, and, potentially, the increased use of Air Force transport aircraft, such as larger numbers of airlift crews trained for airdrop operations.

Of course, all of these options will be driven by how much of the airborne force will be converted to a light armored configuration. The Army has several organizational options for implementing this new airborne capability, including the following:

1. Create a division-level "airborne light armored infantry battalion" that would provide various types and numbers of LAV-II or Stryker brigades on an as-needed basis. This option would be somewhat similar to the Sheridan battalion that was organic to the 82nd Airborne or the amphibious tractor battalion in each Marine Corps division. A key difference would be that, whereas the old Sheridan battalion and the current Marine Corps amphibious tractor battalions are single-mission units with one type of vehicle, this divisional battalion would have multiple vehicle variants: infantry fighting vehicle, reconnaissance, assault gun, and so forth. In this organizational construct, the division-level battalion would have several companies of light armor that would be attached to the airborne infantry brigades and battalions on an as-needed basis. Having sufficient APC versions to motorize one airborne infantry battalion, plus providing an assault gun (the Stryker MGS version or the LAV-II 90-mm assault gun), indirect fire (such as the 81- or 120-mm mortar versions of LAV-II), and reconnaissance (e.g., the

LAV-25A2) capacity would add important capabilities to current airborne units. This versatility could also be integrated with current HMMWV-mounted airborne infantry units. If only one battalion-sized unit was created in the 82nd (and possibly company-sized equivalents in the 173rd and 4th Brigades of the 25th Division) only a limited number of vehicles (probably no more than one company in the case of the 82nd) would be available for short-notice deployment.

2. Convert one battalion in each airborne brigade. This option would allow each airborne brigade to have enhanced capacity in the form of one battalion of airborne light armored infantry, similar to the organizational construct in Chapter Four.

3. Convert one, two, or three airborne infantry brigades to this configuration. This option would provide increased capacity to more of the airborne force. Under this option, one or more airborne brigades would be completely converted to a configuration similar to the Stryker- or LAV-armed brigades in Chapter Four.

4. Convert the entire airborne force. Under this option, all airborne brigades would be motorized and would include light armor. This would clearly be the most ambitious option and would carry the most costs and trade-offs.

In the case of options 1 and 2 where either a single battalion or one battalion per airborne brigade were created, it might be possible to increase the level of motorization of the remaining infantry battalions via the use of HMMWV or similar vehicles. In that case, the airborne light armored battalions would provide a core capability that would operate in conjunction with the less-well-protected but more mobile infantry that would use unarmored vehicles.

Any of these options would have to also be assessed in terms of its impact on the important readiness cycles that airborne units prepare for. For example, if a single division-level LAV or Stryker battalion were created (option 1), it would probably be necessary for an intrabattalion, company-level readiness cycle in that battalion to ensure a portion of the unit was always available to support the infantry brigades. Interestingly, this was done in the old Sheridan battalion in the 82nd Airborne Division, where tank companies rotated being on ready status.

Ultimately, the Army will need to decide whether or not this new airborne concept provides enough additional capability to the force to warrant the costs associated with its development and implementation. The Army will also need to examine the opportunity costs associated with the development of this capability. In other words, what kinds of missions, training, and procurement will be sacrificed if this capability is developed. The size and capabilities of the current and near-future Air Force transport aircraft fleet should be a consideration for any Army airborne organizational option as well.

CHAPTER EIGHT

Conclusions and Recommendations

This report represents the initial step in a multiyear RAND effort to help the Army identify options to enhance its airborne forces, with emphasis on the near term. The research presented in this document highlighted real or perceived weaknesses or vulnerabilities in today's airborne forces, threat trends that are influencing how and under what conditions airborne forces can be used, and possible options for the near future.

A major conclusion of this initial step in the research has been that LAV-II–class vehicles appear to provide an attractive option for Army airborne forces, assuming that a new capability is desired in the three-to-five-year time frame. This does not mean that the LAV-II is the only vehicle option that might be appropriate for use in the manner described in this report. However, if time is indeed of the essence in terms of enhancing airborne forces, using a vehicle that is already part of the U.S. and allied vehicle fleets and compatible with existing airdrop technology and Air Force transport aircraft is an important advantage.

Recommendations

There is a need for the Army to first determine if the light armor concept is right for the airborne force. If this direction appears to be an appropriate one, the next steps should include the following:

- Refine the operational concepts associated with a new airborne light armored infantry capability. This would include a detailed examination of how such a new capability would be employed and what key joint enablers would be necessary for this mode of operations.

- Conduct an experimentation program. This could include obtaining LAV-II–class vehicles from the Marine Corps and elsewhere to examine their suitability for airdrop and air-landing operations. Additionally, Stryker should undergo a detailed assessment to compare it to the LAV-II series and determine its applicability for the airborne mission.

- Examine other vehicle options. This could include determining whether there are any other readily available U.S. or foreign vehicles that might be appropriate.

- Determine what the Air Force's main constraints would be to operationalize some portion of this concept. For example, the Army might desire that more C-17 crews be airdrop-capable than is the case today. The Air Force could not make such a change on short notice. Time and resources would be involved, and the Army and Air Force would need to discuss what is possible under the Army's desired timelines and the proportion of the airborne force that it would want to convert to this configuration. The Air Force transport fleet is important to the entire joint force, so new Army airborne concepts will probably require vetting not only with the Air Force but also through the Joint Requirements Oversight Council process.

- Identify additional rigging and other administrative requirements from the Army's perspective. For example, while there is currently a significant rigging capability at Fort Bragg–Pope Air Force Base in North Carolina, a new requirement to rig several dozen light armored vehicles for rapid deployment could impose a burden beyond the capacity of the current rigging system. For Italy- and Alaska-based airborne units, new rigging capacity and other infrastructure might be needed to accommodate light armored vehicles in the LAV-II or Stryker class.

- Establish the costs for the various vehicle options and the associated new units that would be required (i.e., more rigging capability and maintenance for light armor in airborne units).

- Decide on an initial organizational construct. For instance, does the Army want to convert one or more brigades to this configuration or just a single battalion?

As with any other new military concept, there are possible advantages and disadvantages. Table 8.1 highlights the important aspects of the concept and describes the advantages and disadvantages.

An important step that might come next is presenting the potential usefulness of this enhanced airborne capability to combatant commanders. At present, when theater commanders and their staffs consider incorporating Army airborne forces into their contingency plans, they are basing their decisions on today's airborne forces. Some version of the concept presented in this document would be a new capability. Therefore, the Army should solicit the input of the joint headquarters that would be the ultimate customers and users of this new Army capability.

Table 8.1
Potential Advantages and Disadvantages of New Airborne Concepts

Key Aspect of Enhanced Airborne Concept	Advantages	Disadvantages
Enhanced mobility, protection, and firepower of airborne units	Increases strategic and operational options for airborne forces Tactical flexibility improved	More airlift required to deploy airborne units with larger numbers of vehicles Cost of procuring light armored vehicles for the airborne force
Battalion-sized airborne light armored infantry units	Battalions easier to deploy via existing Air Force airlift assets Fewer vehicles need to be purchased compared to brigade-sized units	Battalion-sized units would probably only be able to maintain company-sized elements on high readiness Limited overall combat power
Brigade-sized airborne light armored infantry units	More combat power than battalions Able to maintain a full battalion at high level of readiness	Would require considerable airlift to deploy Higher cost due to larger number of light armored vehicles that would have to be procured
LAV-II family of vehicles	Well suited to airdrop and transport due to weight and size (LVAD-compliant), including C-130 Family of vehicles already exists, including in U.S. military use Some compatibility with Stryker (same manufacturer)	Not currently an Army system Still-to-be-determined number of vehicles would have to be procured
Stryker family of vehicles	Currently in U.S. Army use Family of vehicles already exists	Currently, Stryker is beyond weight limit of the LVAD system Difficult to transport in C-130, and cannot be dropped from C-130 Additional vehicle types would have to be procured
Use of the current Air Force airlift fleet	No new aircraft purchases needed Air Force familiar with current Army airborne concepts	Additional C-17 aircrew may need to be qualified for armored vehicle airdrop Other elements of the joint force also require airlift, especially from the C-17

LAV-II Family of Vehicles

The Marine Corps currently owns and operates seven different LAV variants. These include the LAV-25 (reconnaissance vehicle), the LAV-C2 (command and control vehicle), the LAV-AT (antitank vehicle), the LAV-LOG (logistics vehicle), the LAV-M (81-mm variant mortar vehicle), the LAV-R (recovery vehicle), and the LAV-MEWSS (mobile electronic warfare system vehicle).[1] While these seven Marine Corps variants are possible procurement options for an LAV-equipped Army airborne force, there are other LAV variants on the market today that offer larger troop-carrying capacity and more firepower than Marine Corps versions.

This appendix provides added detail into the LAV vehicle options available for the Army.

LAV-25

Armed with a M242 25-mm Bushmaster cannon, the LAV-25 (see Figure A.1) is the primary LAV used by Marine Corps light armored reconnaissance battalions. Of the 772 LAVs procured by the Marine Corps from 1984 to 2003, more than half were LAV-25 variants. It

---

[1] The Marine Corps LAV-MEWSS variant is not employed in Marine Corps light armored reconnaissance battalions; rather, it is employed in the two Marine Corps radio battalions and is operated by Electronic Warfare operators instead of traditional 0313 LAV crews. The Marine Corps used to operate the LAV-AD (air defense vehicle) outfitted with Stinger missiles; however, it no longer employs this system.

is operated by three crew members and can carry three or four dismounted infantry personnel. It was built by General Dynamics Land Systems–Canada and is based on the Swiss MOWAG Piranha I 8×8 family of armored fighting vehicles. It weighs 14.1 short tons and has the following dimensions:

- length: 6.39 m (19 ft.)
- width: 2.50 m (8.20 ft.)
- height: 2.69 m (8.83 ft.).

The most advanced version in use by the Marine Corps is the LAV-25A2, which includes upgraded internal and external ballistic protection and an Improved Thermal Sight System that provides the gunner and commander with thermal images, a laser range finder, a fire-control solution, and target location grid information. Reconnaissance versions, such as the Coyote, in Canadian service, can include battlefield surveillance radar systems and other sensors.[2]

LAV-IIH

The LAV-IIH tech demonstrator from General Dynamics Land Systems (see Figure A.2) is an APC variant with an expanded cab that has space for nine dismounted infantry (and two crew members). It is armed with Stryker-like weapon systems (0.50 cal., Mk 19, and M240), has a curb weight of 14.3 tons, and has the following dimensions:

- length: 272 in. (6.9 m)
- width: 104 in. (2.6 m)
- height: 89 in. (2.3 m).

Additional applique armor is available to counter ballistic, artillery, RPG and IED threats. LAV-IIH is General Dynamics Land Systems' current technology demonstrator based on the Piranha II

---

[2] "General Dynamics Land Systems—Canada Light Armoured Vehicle (8×8)," *Jane's Armour and Artillery*, updated November 28, 2011.

Figure A.1
LAV-25

SOURCE: U.S. Marine Corps photo.
RAND *RR309-A.1*

Figure A.2
LAV-IIH

SOURCE: General Dynamics Land Systems promotional image.
RAND *RR309-A.2*

platform; therefore, many additional weapons and upgrades available for Piranha series could be adapted for use.[3]

LAV-AG

The LAV-AG (assault gun version) is currently in use in the Saudi Arabian National Guard (see Figure A.3). It has a Belgian CMI (Cockerill Maintenance and Ingénierie) Mk-8 90-mm gun and a two-person turret; it is built on the LAV-IIH chassis. In 2009, the Saudi Arabian National Guard purchased 84 of these vehicles from General Dynamics Land Systems–Canada.[4]

LAV-AT

The LAV-AT (antitank version), currently in Marine Corps use, is fitted with a first-generation TOW missile system (see Figure A.4). These TOWs have a maximum range of 3,750 m. The vehicle carries two missiles in the ready-to-launch position and 14 additional rounds. A newer version available for purchase from General Dynamics Land Systems has an upgraded LAV-IIH–based chassis. This version is currently in use by the Saudi Arabian National Guard. The Emerson 901 ITV turret used to launch TOW missiles is currently slated for replacement due to its age and the scarcity of replacement parts.[5]

---

[3] General Dynamics Land Systems, "LAV-IIH Technology Demonstrator," web page, undated.

[4] "General Dynamics Land Systems—Canada Light Armoured Vehicle (8×8)," 2011.

[5] "General Dynamics Land Systems—Canada Light Armoured Vehicle (8×8)," 2011.

Figure A.3
LAV-AG

SOURCE: CMI Defence promotional image.

RAND *RR309-A.3*

Figure A.4
LAV-AT

SOURCE: U.S. Marine Corps photo.

RAND *RR309-A.4*

LAV-M

The original LAV-M procured by the Marine Corps is fitted with an 81-mm mortar system that has a 360-degree traverse and can carry more than 90 rounds. The mortar is mounted in the center of the vehicle and fires through the three-part roof hatch. Newer versions, like the one shown in Figure A.5, included the 120-mm variant, which was purchased by the Saudi Arabian National Guard in 2009. This variant features a 120-mm NEMO (NEw MOrtar) system produced by Finnish Patria Land and Armament. The 120-mm NEMO turret has powered elevation and traverse and can be aimed, loaded, and fired while the crew is completely protected. Maximum range of the 120-mm NEMO mortar depends on the munition fired but is claimed to be around 10,000 m.[6]

Figure A.5
LAV-M

SOURCE: U.S. Marine Corps photo.
RAND RR309-A.5

---

[6] "Patria Defense NEMO 120 mm Mortar System," *Jane's Armour and Artillery*, updated February 9, 2012.

Stryker- and LAV-Based Airborne Light Armored Infantry Brigade TOEs

Tables B.1 and B.2 present the TOEs for Stryker- and LAV-based airborne light infantry brigades.

Table B.1
First- and Second-Echelon TOEs

Type	Number						Echelon 1					
		Headquarters and Headquarters Company	Division Headquarters	Infantry Battalion	Cavalry Troop	Engineer Platoon	Engineer Platoon (mobile)	UAV Platoon	Target Acquisition TM and Headquarters TM (Q-36 radar)	Field Artillery Battery	Cavalry Squadron HA	Q-37 Radar
Personnel	1,040	100	30	649	92	14	32	27	90	6		
LAV/Stryker	99	4	1	74	16		4					
M8	9			9								
M777	6									6		
HMMWV	72	17	6	26	1	1	1	7	9	4		
MTV	37	3		12	2		1	1	19			
HEMMT	4					4						
HEMMT trailer	0											
HEMMT trailer bridge	4					4						
1.25-ton trailer	50	17	4	14	1	1	1	7	3	2		

Table B.1—Continued

Type	Number	Headquarters and Headquarters Company	Division Headquarters	Infantry Battalion	Cavalry Troop	Engineer Platoon	Engineer Platoon (mobile)	UAV Platoon	Target Acquisition TM and Headquarters TM (Q-36 radar)	Field Artillery Battery	Cavalry Squadron HA	Q-37 Radar
Echelon 1 (cont.)												
MTV trailer	12	3		1	1		1		1	6		12
Water buffalo	9			8						1		
UAS	4							4				
Q-36 radar	1								1			
Echelon 2												
Personnel	1,137	122		649	32		32		90		140	12
LAV/Stryker	116	4		74	16		4				18	
M-8	9			9	1		1				18	
HMMWV	80	25		26					6		6	3
MTV	34	3		22	2				9		3	2

Table B.1—Continued

Type	Number	Headquarters and Headquarters Company	Division Headquarters	Infantry Battalion	Cavalry Troop	Engineer Platoon	Engineer Platoon (mobile)	UAV Platoon	Target Acquisition TM and TM (headquarters TM and Q-36 radar)	Field Artillery Battery	Cavalry Squadron HA	Q-37 Radar
				Echelon 2 (cont.)								
HEMMT	19									19		
HEMMT trailer	0											
1.25-ton trailer	43	17		14	1		1				9	1
MTV trailer	17	3		1	1				3		8	1
Water buffalo	15			8					6		1	
M1117	7			6					1			
Q-37 radar	1											1

Table B.2
Third- and Fourth-Echelon TOEs

Type	Number	Echelon 3															
		Infantry Battalion	Cavalry Troop	Engineer Platoon (mobile)	Field Artillery Battery	Field Artillery Battalion	Engineer Company Headquarters	Anti-Armor Company	Rest of Field Artillery Slice	Military Intelligence Company	Brigade Support Battalion Headquarters and Headquarters Company	Distribution Company	Office of the Assistant Secretary of the Army for Financial Management and Comptroller	Medical Company	Signals Company	Engineer Rump
Personnel	1,036	649	92	32	90	91	10	53	18							
LAV/Stryker	98	74	16	4		1	1	2								
M8	18	9						9								
HMMWV	67	26	1	1	6	23	2	1	7							
MTV	37	12	2		9	10	2	2								
HEMMT	19				19											
HEMMT trailer	0															
1.25-ton trailer	32	14	1	1		9			7							

Table B.2—Continued

Type	Number	Infantry Battalion	Cavalry Troop	Engineer Platoon (mobile)	Field Artillery Battery	Field Artillery Battalion	Engineer Company	Headquarters	Anti-Armor Company	Rest of Field Artillery Slice	Military Intelligence Company	Brigade Support Battalion Headquarters and Headquarters Company	Distribution Company	Office of the Assistant Secretary of the Army for Financial Management and Comptroller	Medical Company	Signals Company	Engineer Rump
Echelon 3 (cont.)																	
MTV trailer	16	1	1		3			10	1								
Water buffalo	17	8			6			1	1								
M1117	7	6			1			1	1								
Echelon 4																	
Personnel	929										92	153	147	375	89	56	18
LAV/Stryker	3										1					2	
M8	0																
HMMWV	150										23	32	8	38	25	24	
MTV	74										10	17	10	26	10	1	
HEMMT	86																6

Table B.2—Continued

Echelon 4 (cont.)

Type	Number	Infantry Battalion	Cavalry Troop	Engineer Platoon (mobile)	Field Artillery Battery	Field Artillery Battalion	Engineer Company Headquarters	Anti-Armor Company	Rest of Field Artillery Slice	Military Intelligence Company	Brigade Support Battalion Headquarters and Headquarters Company	Distribution Company	Office of the Assistant Secretary of the Army for Financial Management and Comptroller	Medical Company	Signals Company	Engineer Rump
HEMMT trailer	88										12	50	4	10		12
1.25-ton trailer	85									9	21	6	31	17	1	
MTV trailer	31									10	2	9		10		
Water buffalo	11									1	7	1	1	1		
Power plant trailer	5										5					
Containerized kitchen	6										6					
Assault kitchen	15										15					
Forklift	8											7	1			
Bulldozer	6															6
Backhoe	6															6

C-5, C-17, and C-130 Capabilities

In FY 2012, there were 79 C-5s, 364 C-130s, and 180 C-17s in operational squadrons. Each airlift platform could provide benefits to the discussed force options. Table C.1 compares the potential airdrop capabilities of the C-5, C-17, and C-130.

Although the C-5 performed the airdrop mission in the past, it no longer does so. If the C-5 could be recertified for this mission, it could provide significant benefits, especially if JPADS capability encompassed LAV-type vehicle weights. Table C.2 shows the force options using the C-5. The C-5 base and excursion columns assume only C-5s are used, while the combined column employs a mix of C-5s and C-17s.

In addition, the C-5M has greater unrefueled range capability than the C-17 while carrying more airdrop cargo. This additional unrefueled range opens up basing options and decreases the potential reliance on air refueling tankers:

- C-17 carrying two Strykers—2,859 nm
- C-5M carrying three Strykers—3,779 nm.

While the numbers of aircraft required are significantly fewer than that under the C-17-only option, the fact that the C-5 has not accomplished airdrop for many years and there are no qualified crews poses a serious limitation. Medium-altitude airdrop using JPADS may decrease the time required for retraining, but JPADS technology growth must encompass LAV- or Stryker-sized vehicles. This is an area for further study if the Air Force wants to pursue this option.

Table C.1
C-5, C-17, and C-130 Lift Capabilities

	LAV-II		Stryker		Personnel	
Aircraft	Airdrop	Air-Land	Airdrop	Air-Land	Airdrop	Air-Land
C-130J	1[a]	1	0	1	64	92
C-17	3	6	2	3	102	134
C-5	5[b]	8	3–4[b]	4	73[a]	270

[a] The LAV-II has been airdropped by parachute; some versions may be too tall for the C-130.

[b] This is a nominal capability; no C-5 crews are currently trained for airdrop. If variants of Stryker can be rigged under 50,000 lbs, four could be dropped from a C-5.

Table C.2
C-5 Airlift Options

	Number of Aircraft			
			Combined	
Force Option	C-5 Base	C-5 Excursion	C-17 (carrying passengers)	C-5 (carrying vehicles)
Stryker-based airborne brigade (echelons 1 and 2)	186	141	10	141
LAV-based airborne brigade (echelons 1 and 2)	156	111	11	111
Stryker-based airborne battalion task force	45	35	7	35
LAV-based airborne battalion task force	31	21	7	21

Dimensions, Weight, Number of Vehicles for C-17 Airdrop

Table D.1 presents the characteristics of the C-17 airdrop scenario, including the dimensions, weight, number of vehicles, and assumed features of the vehicles.

Table D.1
C-17 Airdrop Scenario Characteristics

Given	Assumed	Length (inches)	Width (inches)	Height (inches)	Weight (lbs)	Number of Airdrops (C-17)	Notes
Stryker	Rigged with level II armor	395	141	113	55,000	2[a]	
LAV	Rigged with level II armor	< 272	103	106	36,500	3[a]	
MTV	M1093 5-ton 6x6 standard cargo truck	354	108	100	27,318	2	
MTV trailer	M1095 cargo MTV	206	108	99.5	30,330	2	
HMMWV	M1114 utility truck, uparmored	222.5	94	106	14,500	8	Dual-Row Airdrop System
HMMWV trailer	M101A1 trailer	216	94	98	8,062	8	Dual-Row Airdrop System
Water buffalo	M149A1 400-gallon water trailer	162	108	86	7,200	6	
M777	M777 howitzer medium towed	383	109.5	94	19,400	2	
M-8	M-8 (level III armor)	350	106	100	49,500	2	
M1117	M1117 armored security vehicle	237	101	102	29,560	4	
M119	M119 howitzer, light, towed	216	94	98	11,200	8	Dual-Row Airdrop System
HMMWV trailer	1.5-ton trailer	166	108	81	7,360	6	
Bridging	Five-bay, single-story bridge	402	108	97	22,480	2	
Bridging	Koehring 7.5-ton crane	347	108	100	30,368	2	

Table D.1—Continued

Given	Assumed	Length (inches)	Width (inches)	Height (inches)	Weight (lbs)	Number of Airdrops (C-17)	Notes
Q-36 radar	TPQ-36V1					0[c]	
	Mounted on M116A1 trailer chassis	183	84	86	6,619		
Q-37 radar	AN/TPQ-37V8					2[a]	
	Operator control, general purpose, mounted on M1097	189	86	104	7,796		
	Power distributor, general purpose, mounted on M925	332	98	121	29,896		
	Antenna transport, general purpose	196	96	92	10,855		
	M1048 for antenna transport, general purpose	235	96	37	5,880		
UAS	Assuming Raven B						
HEMMT	M1977W/CBT	395	141	113	37,240	2[b]	
HEMMT trailer	1076 trailer-plus, 16.5 ton	304	96	124	16,530	2[b]	

SOURCE: Headquarters, U.S. Department of the Army, and U.S. Department of the Air Force, 2006.

[a] Meets airdrop dimension requirements but not currently certified for airdrop.

[b] One or more dimensions do not meet airdrop requirements; would require modifications to certify for airdrop.

[c] Can be packed with other loads.

Bibliography

Air Force Flight Test Center, *Engineering Feasibility Assessment of C-17A Aerial Delivery of the U.S. Army Stryker Mobile Gun System Vehicle*, AFFTC-TR-04-38, November 2004.

Boeing Corporation, "C-17 Globemaster III," fact sheet, February 2014. As of July 22, 2014:
http://www.boeing.com/assets/pdf/defense-space/military/c17/docs/c17_overview.pdf

Collins, Thomas W., "173rd Airborne Brigade in Iraq," *Army Magazine*, June 2003, pp. 42–46. As of July 22, 2014:
http://www.ausa.org/publications/armymagazine/archive/2003/6/Documents/Collins_0603.pdf

Dallaire, Romeo, *Shake Hands with the Devil: The Failure of Humanity in Rwanda*, Toronto, Ont.: Random House Canada, 2003.

D'Angina, James, *LAV-25: The Marine Corps' Light Armored Vehicle*, Oxford, UK: Osprey Publishing Company, 2011.

Defense Science Board Task Force on Mobility, *Report of the Defense Science Board Task Force on Mobility*, Washington, D.C.: Office of the Under Secretary of Defense for Acquisition, Technology, and Logistics, September 2005.

DeVore, Marc, *The Airborne Illusion: Institutions and the Evolution of Postwar Airborne Forces*, working paper, Cambridge, Mass.: Massachusetts Institute of Technology, June 2004.

Feil, Scott R., *Preventing Genocide: How the Early Use of Force Might Have Succeeded in Rwanda*, New York: Carnegie Corporation, 1998.

Fielder, Mark, "The Battle of Arnhem (Operation Market Garden)," *BBC History*, February 2, 2011. As of July 22, 2014:
http://www.bbc.co.uk/history/worldwars/wwtwo/battle_arnhem_01.shtml

Gates, Robert M., Secretary of Defense, transcript of speech at the United States Military Academy, West Point, New York, February 25, 2011.

General Dynamics Land Systems, "LAV-IIH Technology Demonstrator," web page, undated(a). As of July 22, 2014:
http://www.gdls.com/index.php/products/lav-family/lav-ii-h

———, "Overview," web page, undated(b). As of July 22, 2014:
http://www.gdls.com/index.php/products/products-overview

———, "Stryker ICV, DVH," web page, undated(c). As of July 22, 2014:
http://www.gdls.com/index.php/products/stryker-family/stryker-icvdvh

"General Dynamics Land Systems—Canada Light Armoured Vehicle (8×8)," *Jane's Armour and Artillery*, updated November 28, 2011.

Gordon, John, David Johnson, and Peter A. Wilson, "Air Mechanization: An Expensive and Fragile Concept," *Military Review*, January–February 2007, pp. 66–69.

Gordon, John, Peter A. Wilson, Jon Grossman, Dan Deamon, Mark Edwards, Darryl Lenhardt, Daniel M. Norton, and William Sollfrey, *Assessment of Navy Heavy-Lift Aircraft Options*, Santa Monica, Calif.: RAND Corporation, DB-472-NAVY, 2005. As of July 22, 2014:
http://www.rand.org/pubs/documented_briefings/DB472.html

Gourevitch, Philip, *We Wish to Inform You That Tomorrow We Will Be Killed with Our Families: Stories from Rwanda*, New York: Picador, 1999.

Grant, Greg, "A Little Bird for the Army?" *DoD Buzz*, October 23, 2008. As of July 22, 2014:
http://www.dodbuzz.com/2008/10/23/a-little-bird-for-the-army

Harrison, Gordon A., *Cross-Channel Attack*, Washington, D.C.: U.S. Army Center of Military History, 1951.

Headquarters, U.S. Department of the Army, *The Soviet Army: Specialized Warfare and Rear Area Support*, Field Manual 100-2-2, Washington, D.C., July 16, 1984.

———, *The Stryker Brigade Combat Team Infantry Battalion*, Field Manual 3-21.21, Washington, D.C., April 8, 2003.

———, *Reconnaissance and Cavalry Troop*, Field Manual 3-20.971, Washington, D.C., August 4, 2009.

Headquarters, U.S. Department of the Army, and U.S. Department of the Air Force, *Airdrop of Supplies and Equipment: Reference Data for Airdrop Platform Loads*, Field Manual 4-20.116/Air Force Technical Order 13C7-1-13, Washington, D.C., May 10, 2006.

Hoffman, Frank G., "Hybrid Warfare and Challenges," *Joint Force Quarterly*, No. 52, 2009, pp. 34–39.

International Commission on Missing Persons, "Over 7,000 Srebrenica Victims Have Now Been Recovered," July 11, 2012. As of July 22, 2014:
http://www.ic-mp.org/press-releases/over-7000-srebrenica-victims-recovered

International Institute for Strategic Studies, *The Military Balance*, 2012, London, March 2012.

Johnson, MG James H., Jr., interview with Robert K. Wright, Jr., historian, XVIII Airborne Corps, at Headquarters, 82nd Airborne Division, Fort Bragg, N.C., March 5, 1990. As of July 22, 2014:
http://www.history.army.mil/documents/panama/JCIT/JCIT26.htm

Kopp, Carlo, "Are Helicopters Vulnerable?" *Australian Aviation*, March 2005, pp. 59–63. As of July 22, 2014:
http://www.ausairpower.net/PDF-A/TE-Helo-Mar-05-P.pdf

Krump, Jamie L., "Sustaining Northern Iraq," *Army Sustainment*, November–December 2003. As of July 22, 2014:
http://www.almc.army.mil/alog/issues/NovDec03/Sustaining_Northern_Iraq.htm

"LAV-Assault Gun," *Army Guide*, undated. As of July 22, 2014:
http://www.army-guide.com/eng/product971.html

Longabaugh, MAJ Raymond, U.S. Army G-4, 82nd Airborne Division, interview with the authors, September 11, 2012.

Matsumura, John, Randall Steeb, John Gordon, Thomas J. Herbert, Russell W. Glenn, and Paul Steinberg, *Lightning Over Water: Sharpening America's Light Forces for Rapid Reaction Missions*, Santa Monica, Calif.: RAND Corporation, MR-1196-A/OSD, 2000. As of July 22, 2014:
http://www.rand.org/pubs/monograph_reports/MR1196.html

McKeever, Maj Scott, SAF/IARA, interview with the authors, RAND Corporation, Arlington, Va., July 24, 2012.

Nardulli, Bruce R., Walter L. Perry, Bruce R. Pirnie, John Gordon, and John G. McGinn, *Disjointed War: Military Operations in Kosovo, 1999*, Santa Monica, Calif.: RAND Corporation, MR-1406-A, 2002. As of July 22, 2014:
http://www.rand.org/pubs/monograph_reports/MR1406.html

"News Archives: Chinese Missile Defenses," *Missile Threat*, undated.

Obama, President Barack, "Remarks by the President on the Defense Strategic Review," transcript, January 5, 2012. As of July 22, 2014:
http://www.whitehouse.gov/the-press-office/2012/01/05/
remarks-president-defense-strategic-review

O'Rourke, Maj Joseph J., AF/A8XS, interview with the authors, RAND Corporation, Arlington, Va., July 24, 2012.

"Patria Defense NEMO 120 mm Mortar System," *Jane's Armour and Artillery*, updated February 9, 2012.

Ryan, Cornelius, *A Bridge Too Far*, New York: Simon and Schuster, 1974.

"Saudi Arabia Orders LAV IIs for National Guard," *Jane's Defence Weekly*, December 24, 2009.

Scales, Robert H., Jr., *Certain Victory: The U.S. Army in the Gulf War*, Washington, D.C.: Potomac Books, 2006.

Shlapak, David A., John Stillion, Olga Oliker, and Tanya Charlick-Paley, *A Global Access Strategy for the U.S. Air Force*, Santa Monica, Calif.: RAND Corporation, MR-1216-AF, 2002. As of July 22, 2014:
http://www.rand.org/pubs/monograph_reports/MR1216.html

Srebrenica-Potocari Memorial Center, list of victims, undated. As of July 22, 2014:
http://www.potocarimc.ba/_ba/liste/nestali_a.php

Thornton, Rod, *Organizational Changes in the Russian Airborne Forces: The Lessons of the Georgian Conflict*, Carlisle, Pa.: Strategic Studies Institute, U.S. Army War College, December 2011. As of July 22, 2014:
http://www.strategicstudiesinstitute.army.mil/pdffiles/PUB1096.pdf

Twiford, Lt Col James R., AF/A5RM, interview with the authors, RAND Corporation, Arlington, Va., July 24, 2012.

United Nations, "'May We All Learn and Act on the Lessons of Srebrenica' Says Secretary-General, in Message to Anniversary Ceremony," transcript of speech by United Nations Secretary-General Kofi Annan delivered by Mark Malloch Brown, United Nations Chef de Cabinet, Potocari-Srebrenica, Bosnia and Herzegovina, Press Release SG/SM/9993, July 11, 2005. As of July 22, 2014:
http://www.un.org/News/Press/docs/2005/sgsm9993.doc.htm

U.S. Air Force, "C-17 Globemaster III," fact sheet, October 27, 2004. As of July 15, 2014:
http://www.af.mil/AboutUs/FactSheets/Display/tabid/224/Article/104523/c-17-globemaster-iii.aspx.

———, "C-130 Hercules," fact sheet, May 2014. As of July 22, 2014:
http://www.af.mil/AboutUs/FactSheets/Display/tabid/224/Article/104517/c-130-hercules.aspx

"Weishi (WS-1/-2) Multiple Launch Rocket Systems," web page, last updated December 31, 2008. No longer available.